PRAISE FOR *PRESSING ON AS A TECH MOM*

"As a global technology leader and mother of two boys, I can attest to the challenges moms face in the workforce. The stories D'Anzica and Pons share illustrate how you can be a high performer regardless of gender and take the time to be there for your family, too."

-Catherine Blackmore, General VP of NAA Customer Success and Renewals at Oracle

"Through their stories and advice for women navigating their careers in technology companies today, D'Anzica and Pons drive to the heart of the challenges for working mothers in tech."

-Alana Karen, Director of Search Platforms at Google and author of *The Adventures of Women in Tech: How We Got Here and Why We Stay*

"As a tech mom who painstakingly birthed my first during a pandemic, experienced a subsequent miscarriage shortly after and is close to birthing my second, the authentic stories in this book resonate deeply. *Pressing ON as a Tech Mom* is a therapeutic, inspiring and critical compilation of subsumed female experiences that need to be socialized and internalized at the highest levels of tech leadership to ensure the current progress continues to flow in a forward-facing direction."

-Samma Hafeez, VP of Sales and Customer Success Center of Excellence at Insight Partners

"I didn't have any female role models or mentors when studying at a technical university, working in venture capital or first becoming a tech CEO. Fortunately, D'Anzica and Pons are changing all that for future generations by sharing the real-life experiences of other tech moms."

-Dominique Levin, CEO of Winning by Design

"Twenty-five years in, I'm still learning how to be a mom in tech. While there is no one answer for all of us, D'Anzica and Pons take us on a journey of awareness and most importantly, acceptance."

-Giorgia Ortiz, VP of Strategy and Revenue Enablement at Jellysmack

"This book will be my new go-to gift for fellow moms in technology. *Pressing ON as a Tech Mom* provides an unprecedented view into the struggles of mothers in the tech world *and* gives tips on how to succeed."

-Nikki Bishop, VP of Customer Success at Seeq Corporation

"*Pressing ON as a Tech Mom* is useful on many levels and contains a message that resonates: women can join tech, stay in tech and thrive in tech without foregoing motherhood."

-Kellie Capote, CCO of Gainsight

"A beautiful reminder of the necessity of sister-hood and sponsorship in business."

-Jeanne Bliss, Founder and CEO of Customer Bliss, global keynote speaker and bestselling author of *Chief Customer Officer 2.0*

"At Gainsight, we are constantly exploring ways to promote an inclusive culture for all of our teammates in line with our 'human-first' philosophy. I am so grateful to D'Anzica and Pons for sharing their wisdom and inspiring so many with this important book."

-Nick Mehta, CEO of Gainsight

"*Pressing ON as a Tech Mom* offers insights, tips and stories of women who have shown how mothers can succeed in a previously male-dominated space. If you're looking for inspira-tion, this book is for you."

-Jim Wilson, Operating Partner at Costanoa Ventures

"For women in tech navigating motherhood: this book is for you. The insights that D'Anzica and Pons have captured surely would have guided me in my earlier career in technology."

"At Industrial Logic, we pride ourselves on diversity and inclusion. As supporters of and mentors to mothers in technology, and as leaders, we must be responsible for creating ways to ensure moms are a critical part of the tech success story."

PRESSING ON

AS A

TECH MOM

HOW TECH INDUSTRY MOTHERS
SET GOALS, DEFINE BOUNDARIES
& RAISE THE BAR FOR SUCCESS

EMILIA M. D'ANZICA
& SABINA M. PONS

AUTHORS' NOTE

This work is non-fiction and, as such, reflects the authors' memory of the experiences. Many of the names and identifying characteristics of the individuals featured in this book have been changed to protect their privacy, and certain individuals are composites. Dialogue and events have been recreated; in some cases, conversations were edited to convey their substance rather than written exactly as they occurred.

CONTENTS

We dedicate this book to our children:

Sawyer, Savannah, Ava, Elle and Tyler.

Vi ameremo sempre

We also dedicate this book to all working moms, people thinking of becoming moms and the leaders who elevate mothers in technology.

"If you want to lift up humanity, empower women."

MELINDA GATES, *THE MOMENT OF LIFT*

PART 1
INTRODUCTION

ONE

WOMEN IN TECHNOLOGY

> "Humans are allergic to change. They love to say, 'We've always done it this way.' I try to fight that. That's why I have a clock on my wall that runs counter-clockwise."
>
> —Grace Hopper, computer programming pioneer

Embracing the Feminine Instinct to Thrive: Emilia's Story

I grew up in a family of eight, in a pizzeria in Campania, Italy until the earthquake of Irpinia.[1] The 6.9 earthquake shook my family with fear, and in 1981, we began our immigration journey to Kelowna, British Columbia, Canada.

Our passion for pizza continued in Canada, where

my family created customer experiences that brought back repeat customers, referrals and newspaper features. In those formative years, we were surrounded by restaurant patrons who not only feasted on my parents' labors of love, but also supported my family for over 30 years. The experience set the stage for me to dedicate my professional life to serving others. Never did I imagine my family's livelihood would lead me to building a career in Silicon Valley in project management and customer success. The lessons I learned cleaning tables, serving pizza, working for tips, seeing the smiles on those familiar faces and having so much fun with my family are the foundation of my passion for helping mothers in tech thrive by teaching them to create their own financial opportunities and happiness. Rather than feeling like a chore, work feels joyful when I help others succeed.

My sisters and brothers worked equally hard, as did my mom and dad as partners in business. Now in their mid-80s, my parents lack understanding of what I do, but when I tell them my profession is similar to what I grew up doing as a child, they laugh and question my pizza-making skills.

My very first foray into tech was perhaps the best decision of my life, and it began with a passion—a passion to change the world. My first mentor and boss, Casey Seidenberg, hired me for my first tech job at the age of 22, as a newcomer to San Francisco who did not know a thing about software as a service (SaaS). Casey is a strong woman who helped me find my footing in the field, but as I began moving up the leadership hierarchy, the adversity I faced in the workforce increased. Examples of this ranged from workplace harassment to repeated colleague scrutiny. One guy on my team publicly called me out for leaving at 5 pm daily to pick up my children at daycare, complaining he was putting in longer hours than me. As I navigated my career, I felt attacked for being a female and constantly found myself defending my actions. As a mother, the struggles only intensified. I remember building up the courage to tell my boss that I did not want to travel to client meetings because I wanted to continue breast-feeding my child. Compounding my mental stress was the fact that I had to muster up the courage to tell clients that I needed to take a break every two hours to pump milk with a hand pump, preferably in a room with a lock as opposed to a toilet stall.

These uncomfortable experiences often made me question if tech was really the right place to be. These conversations were awkward and filled me with anxiety as a new mother. I did not know how to handle them and feared the worst outcomes. Only years later did I realize I was not the only woman facing the same dilemmas. Despite the setbacks and challenges, I wondered, *How do I continue this career that I am so passionate about?*

I sometimes felt that doing it on my own was impossible, but as I looked around me, I realized I was not alone. I saw tech leaders who were working mothers stepping up and embracing the challenges all while reshaping the tech world. We were effectively questioning what was normal and showing that moms could be part of the Silicon Valley success story. As I read more blogs about moms in technology and talked to more mothers in my community, I felt inspired. I knew that if they could do it, so could I.

Soon after I had my second child, I questioned again whether I could keep up the grind. One woman I met while working at BrightEdge was Lorna Henri, a Stanford engineering and MBA graduate, tech leader in the San Francisco Bay Area and an amazing mother of three boys. She became one of my role models who reminded me through her daily actions that I deserved to have a say in how companies were being shaped. I was hungry to learn how to be an inspiring and balanced mother while being a technology professional. In this knowledge quest, I discovered authors like Sheryl Sandberg, Emily Chang and Alana Karen, who have written books about their journeys as women within the tech world. Their stories and research gave me further strength to speak up about why and how mothers belonged in the tech workplace.

The stories they share in their books are at times unbelievable, often upsetting and yet always inspiring. They are true stories giving validation to the slow but improving support mothers get in the workplace. The challenges these authors describe for mothers, women and especially non-white women existed everywhere, but were rarely discussed before books like theirs. They have helped women in general to know that their voices are heard, their struggles are real and that they can change the world. As I grew in my career, I was also blessed to find mothers who shared my struggles, and together, we stood up for each other and rose above our challenges.

Throughout my journey in tech, there have been powerful women who empowered me along the way. All of my thanks must first go to my mom, Margherita, who reminded me continuously that I could do anything that I put my mind to. I attribute my grit in overcoming hardship to her. She experienced the grief of a lost life when my brother died at a young age after getting hit by a drunk driver. In spite of her pain, my mother had the courage to pick herself up and stand strong for the rest of her children. Even when my dad was not supportive of my goal to continue my education in a university, my mother was the one who handed me money under the table and motivated me to pursue my passion. She was the one who pushed me to keep moving forward and did everything in her power to assist in my journey.

Along with my mother, my older sister, Josephine, also introduced me to the power of education and even moved to California with me so we could start our careers in the same company. She has stuck with me throughout the journey and stood by me for every challenge and difficulty. I've also learned a lot from her own experiences and have been there to support her during her own struggles.

While we often discuss the misogyny within the tech world, there are several men who have also supported me throughout my journey. The most prominent is my partner, Jason, who has championed me for standing up for myself. Even if he did not agree with what I was doing, he helped me role-play difficult conversations that I dreaded and supported me at every point in my journey.

Nineteen years into my career, I reached a point where I was mentoring other women and mothers in their own careers. It had been my passion from the start, and to realize that achievement gave me a sense of fulfillment. It was during that time that I met Sabina during a workshop at her company. She confided in me almost immediately and told me that she was struggling as she tried to balance being a mother with leading a global technology organization. I gave her my utmost support, and that's how our friendship started. She reached out to me again last year during the pandemic, and that is when we decided to write this book and to work together in a more formal way. It was empowering for both of us to help others with their careers as mothers in tech.

I'm amazed at how far Sabina and I have come with our company Growth Molecules, particularly when looking back to 1999, when I started my career in San Francisco at Guru.com. As I look at the tech industry today, I am overjoyed to see so much support available for working mothers. I feel that we are going in the right direction now. One of Growth Molecules' customers, Lisa Graham, is a working mother who was recently promoted to CEO of a SaaS engineering company, Seeq. Seeing that kind of growth continually inspires us to keep helping working mothers rise in their passions. In the future, I aspire to see mothers dominating the tech sphere.

Despite this progress, there are still deeply rooted issues within tech that need to be removed. There needs to be a greater degree of empathy and flexibility in the workplace so mothers can balance their personal lives along with their careers. The onset of the COVID-19 pandemic in 2020 taught me the importance of these aspects on a personal level. I have taken the time to become more empathetic while being more vocal about my struggles as a working mother. I have incorporated my personal life more openly in my career as well, and while working from home, I have been thankful to be able to spend more time with my children (and my chickens, cat and dog!).

I am personally satisfied with my place in tech now, and my advice to mothers everywhere is to stand up for themselves and never give up. Tech is changing, and there are so many opportunities for aspiring tech mothers. Now is the time to seize and realize your dreams.

—Emilia

My Why for This Book

I regularly witness how challenging it is for women, with or without children, to openly express themselves without fear of assault in the world. This reality inspires me to support and motivate other women in their journey through tech. I developed a set of goals to achieve my personal ambitions early in my career, and the first step to realizing them was to remove the stigma and fear that women often associate with tech. I want to illustrate through my actions that moms undoubtedly belong in tech, free of persecution and doubt of ability.

I have personally witnessed the struggles that mothers have had to go through from harassment, belittlement and financial discrimination. Despite the struggles, so many women have proven that it is possible to be a successful working mom. The CEO of HubSpot, Yamini Rangan, is taking the stock to new heights; Sara Blakely, who is leading one of the fastest-growing firms, is revolutionizing how and what we wear; and all the women who pick themselves up each time their ideas are rejected and try again are inspiring others to do the same.

With a rising number of mom role models in tech, the question in my mind always toggled between "Why do women still face so much discrimination working in tech?" and "Why are there so few women on the top 100 technology leadership lists?" After much deliberating, I believe the biggest reason is a lack of awareness regarding the achievements of mothers in the tech world. This book serves to bring those stories to light and inspire other mothers to step up and show their capabilities to the world. Our book offers guidance and support and creates the feeling necessary for mothers to succeed: hope.

I want to give women the confidence they need to stand out. I want to show moms that there is a way to continue a career in tech after having children, even if you leave for several years to care for a loved one. In the future, I want these women to be leading tech from the front lines, so they may reshape the world into a more inclusive place where everyone can excel.

A Woman's Influences and Comparisons: Sabina's Story

In adolescence, my grandmother and mother often told me the motivational words that I could be anything I wanted to be, and that I had "it." I think they meant that I could write my own life narrative, and that I had charisma to assist along the way. This was comforting to hear but also daunting. As the eldest child among my siblings and cousins, I felt pressure to be exemplary in all that I did, and that was over-whelming at times.

I was born in the early 1980s, and by the end of the decade, I wore permed hair, scrunched socks and acid-washed jeans as most young girls did. I filled my head with mainstream pop songs like Janet Jackson's chart-topping "Control." *The Atlantic* said in 2014 that her hit "announced the arrival of a young woman ready to take the reins of her personal life and career." I heard you, Janet, and I was also hungry for independent success. [2]

As the daughter of baby boomers, my parents pres-sured me to work hard, because they believed in the promise that hard work contributed to future success. Baby boomers are characterized by the post-war era in which they were born (1946-1964) and are known for striving for the "American Dream."[3] This translated to my mother, Peggy, being among the women pioneers who achieved unprecedented professional success. Her ambition and confidence certainly contributed to mine and were underscored by words of encouragement and gestures of support.

My mom was an executive for a national staffing agency, and then an executive for the fundraising arm of a large regional hospital. She outranked most of her female colleagues. I watched her get ready each morning, as young daughters often do with their mothers. She curled her bangs to the right poofy height and pulled the rest of her hair back into a low ponytail, secured by a large bow on a clip (she, too, fell victim to the egregious 1980s trends). Her lips were swiped with an approachable shade of berry wine to complement her olive Sicilian skin tone. She dressed in shoulder-padded power dresses with clean lines that gave a nod to nautical fashion, of which she is still fond today. On went the corporate-dress-code-required pantyhose, the modest black leather, closed-toe pumps and a spritz of Giorgio Beverly Hills perfume before she ushered us out the door to drop us off at the bus stop on her way to work.

The COVID-19 pandemic of 2020 aside, the morning routines of working moms remain similar in some ways. Thankfully, gone are the days of pantyhose and frizzy bangs. The headspace required to pack lunches, walk the dog, grab backpacks and do QA on the kids' tooth brushing is virtually the same as it was in the 1950s; the main difference between what my maternal figures experienced and what my generation of working moms experience is technology.

Technology has enabled a follow-the-sun philosophy, a global workflow where issues pass between different international offices and time zones as the sun rises in each location. This is great for technical client support teams who are staffed accordingly and horrendous for working moms who lack a capacity model that suggests alternate resourcing when they are at maximum capacity. Today's working moms in technology are never really "off." We do all of the aforementioned dog, lunch and kid duties, but now we also have to pack face masks (!!!)—and we have to do it all while listening to a daily web conference call with our London team on hands-free wireless earbuds. My mom occasionally brought the stresses of work home with her in the form of printed reports or yellow-lined notepads, carried in her cognac leather attaché case, but even if she attempted to do work before setting into our bedtime rituals, she was usually unsuccessful because my father was mostly absent and my siblings and I lacked the digital babysitters of tablets and gaming consoles my kids have today (no mom shaming allowed!). It meant that by necessity, she could still proverbially power off her work—even if she could not shut down her brain (can any mom really do that anyway?).

—Sabina

My Why for This Book

When my mom, Oma and Opa told me I could do anything, I believed them, and most importantly, I believed in myself. When I made it to university, I was beaming with pride and optimism for all that I knew I could contribute. It was not until I hit the workforce in the early 2000s that I saw something was off. I started to notice how my female superiors at work were regarded by their male counterparts when the women had to leave to pick up their kids from daycare. I began to feel the pressures of physical appearance relative to job opportunities. I caught glimpses into calendars of all-male post-work outings. I brushed it off in favor of seeing the glass half full.

As my career progressed, I, too, found myself subject to unfavorable treatment: I had a boss throw pencils at my face when he was angry, and HR did nothing about it after I reported it multiple times ("He just has a bad temper. He does not mean it."). On the tail end of having a miscarriage, instead of resting at home, I was back at work right away despite excruciating bleeding and cramping for fear of being over-looked for an upcoming opportunity for a promotion. When I was pregnant, I vomited every day, multiple times a day, yet I was still expected to deliver optimal outputs. I had to rinse my mouth, wipe my eyes and walk on stage to present live to my entire company, just as I would have any other day.

Postpartum, my male CEO actually told me that I should learn to prioritize my physical fitness as it would give me more stamina for "peak performance" in the workplace. A few months later, I was told that I was overlooked for a promotion because I had fewer years of experience. When I challenged this inaccuracy and highlighted my résumé, I was told by my superior, "Oh! I did not realize that. He looked older. I will have to keep that in mind for next time." So because of my commitment to my skincare regimen to try to help hide the effects of sleepless nights nursing my baby, my bald male colleague was lauded for his years of applied expertise?! Unbelievable! Similarly frustrating was when a file with employee names and salaries was erroneously emailed to the entire company. Aside from the fact that this was unprofessional, what was even more flabbergasting was what the contents revealed: I had become a statistic.[4] I was one of the few women leaders in my company, and all six of us were paid 80 cents for every dollar paid to our male counterparts. This occurred as I was in the thick (literally and figuratively) of working through severe postpartum depression and anxiety while managing a global team of nearly 100 people who were responsible for 70 percent of the company's revenue. When I saw this pay differential, especially relative to the expanded scope of responsibility my female colleagues had compared to their male colleagues, I was absolutely crushed.

These horrible experiences from my career were admittedly partially suppressed and slowly emerged throughout the writing of this book. Mentioning the book project to other women has also resurrected stories from their pasts. In fact, my mom recently shared an infuriating story with me more than 20 years since it happened.

"When I was working in the late 1980s and 1990s in corporate America, women were the absolute worst to one another," my mom said, sharing a story of how her female boss refused to allow her to take time away from work to care for me and my sister when we had chickenpox. "Although my boss was also a mother to two children, she was not maternal. She could not relate to me needing the time off, and instead, my boss sent her nanny to my house to watch you and your sister. The nanny could not speak English and I had to leave you and your sister, then eight and four years old, respectively. You two were both scared and cried. So did I." My mom explained that if she had stayed home, she would have lost her job, and that she felt so torn as the primary breadwinner for our family. Creating a culture where women *support* women, rather than hold them back, is one such way that we can ignite change. Hearing my mom share this story with me reignited my gratitude for the example she set and the sacrifices she made (though she should not have had to).

Besides my mom, another incredible influence in my life is Emilia D'Anzica, my mentor, friend, business partner and co-author of this book. Emilia's care and nurturing of my career is an example of how women *should* treat one another. She and I met just four years ago when she was a management consultant for my corporate team. She was also a strong, confident and positive working mom in technology. I wanted to be just like her and asked if she would mentor me. She said yes, and since then, she has paved the way for me to have public speaking engagements, business oppor-tunities and a new career path when I joined her as a consultant at Growth Molecules in early 2021.

One such way that Emilia and I have found our way to stay in tech is through kinship with other women in tech. *PiMothers*, a website for mothers in tech, summarized this concept well: "Having role models to look up to and learn from helps women make informed choices and gives them hope that they can do it, too. When mothers share their personal stories and experience, it creates awareness towards overall systemic issues around processes and culture in society." I am thankful for Emilia and the other women who have supported my career (Ruth Haiduc, Jodie Davies, Adrienne Ainbinder, Jennifer Chapman and more) and I believe that others can have this opportunity as well.

My reason for co-writing this book is to encourage moms to join tech, stay in tech and thrive in tech, all while supporting one another. It is hard, but it will get better. There are more of us working moms in technology now than ever before. It can be hard and frustrating, but trust me, the journey is sweet and the rewards can be great. While my children do not yet know the stories of my past, they do know the woman that I am today. I am proud of the challenges I have overcome and that I have asked for help from my allies along the way. I am passionate about what I have done and what I can do, and I am going to fight like hell to make the world better for my daughter.

TWO
THE BOOK'S MISSION

Technology has evolved rapidly over the past few decades, with advancements being made in every walk of life. However, the advancements of better treatment and opportunity for underrepresented groups have developed more slowly. It is perplexing that while tech is such an essential part of the world around us, the sector continues to retain its white male dominance despite efforts to make it more inclusive for women.

Our book's main focus is exploring the relationship between motherhood and a career in technology. For this purpose, we have personally interviewed over a dozen mothers in technology and also surveyed over 300 different mothers in this business vertical from around the globe. It was our aim to understand their experiences and how they have overcome challenges to establish successful careers in this competitive space.

Before we delve into the personal accounts of these individuals, it is necessary to understand the current position of women in the tech sector and how this extends to mothers within the tech industry. A key factor to understand is that we are accounting for all the mothers who play a critical and direct role in the tech sector. This includes mothers in management, legal counsel, product management, engineering, finance, professional services, operations, marketing, customer success, sales and all other roles that are crucial for the enablement of technology companies.

Women have often played a crucial role in technology, yet it is alarming to note that the total percentage of women in the technology labor force was only 28 percent in 2020.[1] For every woman employed within technology-related fields, there are more than three times as many men in similar positions. This is already a concerning statistic, but as we move up the corporate ladder, we can see that the position of women weakens more. Only 19 percent of board member positions within technology companies are occupied by women. Furthermore, research shows that for every 97 males employed as CEOs of a company, their female counterparts occupy only three positions. Said differently: Just three percent of the highest executive positions in a company are occupied by women, according to

"Women in Tech Statistics: What the Numbers Tell Us."[2]

While these statistics may be a result of sexism within the technology industry, the problem is much more deeply rooted in other factors. The very first hurdle that limits women's involvement within the tech sector is the commonly held perception that tech subjects are typically a man's forte.[3] This stereotype has often crushed the dreams of aspiring women in technology before they have even started their journeys in the industry. Unfortunately, this is only the tip of the iceberg, as media and news outlets have also contributed heavily to the suppression of women-led achievements within tech. It is common to see women being deprived of the credit they deserve. One of the most prominent examples of this involves the development, engineering and programming of the first electric computer. The Electrical Numerical Integrator and Computer, developed in the World War II-era, was the first of its kind, and the brains behind its programming were six talented women. Additionally, multiple women were crucial to the engineering and construction of the computer. These women were the main driving forces behind the project. Ironically, in the media publication regarding the achievement, these women were referred to as "models" and were equated to nothing more than poster girls.[4] Events like these are both upsetting and discouraging. They do not promote the idea that women should pursue careers in the tech industry. The continuous suppression of women-held achievements has also led to the reinforcement of the belief that only men can be successful in the tech sector. These and other factors have contributed to women's low interest in tech subjects as a whole. As Emily Chang put it in her book, *Brotopia,* "Silicon Valley is a modern utopia where anyone can change the world. Unless you're a woman."

According to a survey by PwC, only 27 percent of women were willing to pursue a career in technology, as opposed to 62 percent of men.[5] This unwillingness has been further exacerbated by the limited number of prominent women, and thus prominent role models, in the technology sector. In one of Growth Molecules' Customer Success Certification classes, one woman recently said in front of all her peers, "I was majoring in math in college until I looked around the room and did not feel like I fit in. I switched majors." People laughed, but it was not a laughing matter. It was an upsetting experience to know a female mathematician left her passion because she did not feel that she fit in. It confirmed that the systemic challenges in science, technology, engineering and math (STEM) fields that women face are rooted early in academia. It is time to change this.

While a majority of women are already hesitant to pursue careers in tech, even those with a vested interest in STEM subjects often become victims of discrimination and sexism. This has proven detrimental to the progression of the careers of women in technology, as companies are typically opposed to awarding women higher positions over their male counterparts. According to Entelo, women only have a 19 percent share within entry and mid-level jobs in STEM fields, and the number drops to 16 percent when considering senior-level positions. Women only occupy 10 percent of executive positions, which means that they are often left out of the essential decision-making process.[6] This low representation has limited the voices of women in tech and made it harder for them to fight against discriminatory policies.

With respect to motherhood, a *HackerRank* survey discovered that 20.4 percent of all women in the industry over the age of 35 remained in junior positions while the percentage for men was only 5.9.[7] Typically, these are the ages of motherhood, and therefore, we can deduce that most of the women being denied these better positions are mothers.

Even when women overcome all the challenges against them, they still have to survive in a male-dominated workspace. Women are already at a disadvantage in terms of social standing due to the inherent bias against us. Sexist attitudes and locker-room culture within and outside tech only make matters worse.

Biological factors can also be a major setback for women in tech. In a grueling industry where mantras of "Better, faster, more, now!" are the norm and consistent work and more work hours are expected, the realities of womanhood and parenthood are neglected. Feeling menstrual cramps, the biological equivalent of a raging war in one's uterus, while preparing a board report, presenting at an all-company meeting or crushing a client pitch—all while keeping a pleasant disposition, staying impeccably groomed and femininely dressed (but not too femininely), as you realize you remembered your youngest had "wacky hair day" at school and that you also forgot to charge your eldest's Chromebook for his web-based standardized testing day—is flat-out overwhelming. It is exhausting, and unfortunately, it is a continuation of other biological disparities between women and men.

Women going through fertility treatments, who are pregnant or who are going through postpartum recovery are all impacted by the physiological and psychological effects of hormones. Medical leaves are often necessary, yet either come with little to no pay, or are provided with pay but also with career setbacks. All too often, women are forced to choose between self-care and career drive. The decision criteria can significantly impact career prospects for both potential mothers and those who already shoulder the responsibilities of motherhood. The good news is that large global tech companies like Netflix, Adobe, Google and Facebook have actively worked to normalize and encourage parental leave, paving the way for other companies to follow.[8] Ironically, some of these same companies, like Google,[9] have been subject to the #MeToo movement, where the realities about the mistreatment of women have been illuminated. No longer can tech companies turn a blind eye. They are part of the problem and they must be part of the solution.

Additional glimmers of hope we can all cling to are the increasing, albeit slowly, numbers of prominent female executives in tech like Sheryl Sandberg (Facebook), Whitney Wolfe Herd (Bumble) and Ginni Rometty (IBM). Each of these women has a story about how they reached these new heights. These stories have been reported in books published about women in technology, among them *She Persisted* by Pratima Rao Gluckman. These works have been influential in our own careers and research for this book. However, we believe there is an opportunity to dive deeper, to look at *mothers* in technology, rather than just *women* in technology. Therefore, this book aims to highlight the challenges of mothers in technology, sharing stories of how they overcome their hardships so that they might pave the way forward for their sisters, daughters and friends. The stories of the mothers in this book will offer a source of guidance and inspiration for other mothers and help them find their place within the tech sector. The goal is to let women and mothers everywhere know that they are not alone, and that they are just as capable as anyone else of being the next successful leaders in technology. The story of Susan Wojcicki, the CEO of YouTube and one of the most influential technology leaders in the global industry, underscores the possibility of success in both career and family life. Susan has revolutionized what it means to be a strong woman in technology through her focus on women's empowerment, all while raising five children.

We hope that mothers everywhere will take inspiration from this book and show the world that a woman can have simultaneous successes, both as mothers and tech professionals. Join us on our journey as we discover the mothers who have struggled, persevered and overcome. Today, we are the minority, but tomorrow, we will be equals.

THREE

A NOTE ON BREAKING IMPOSTER SYNDROME

Before introducing the stories of the women who are the true subject of this book, it may help to first start with the universal before narrowing to the particular. In the case of ambitious women pursuing careers in tech, recognition of a common and relatable problem called *imposter syndrome* should be addressed.

According to the *Harvard Business Review*, imposter syndrome is "doubting your abilities and feeling like a fraud at work," and is something women often experience.[1] The concept was developed in the 1970s and still has quite a hold on many of us in the working world.

When we feel like a fraud, it can keep us frozen in fear so that we do not take action on our goals or work projects. How can we be afraid and do it anyway? Does "fake it until you make it" really work?

As a management consultant and former chief customer officer, one of the book's authors, Sabina, has worked with men and women seeking to grow in their careers but who were holding themselves back because they did not believe in themselves, were embarrassed by their accomplishments or did not want to outshine their bosses. As it is an issue that many readers can relate to, it may be best to allow her to explain her approach to the problem in her own words:

* * *

One of the key principles I use to lead my team is to get comfortable with being uncomfortable. This is how we grow and learn. What works with "faking it" is the skill and confidence that come with practicing something new as we stretch into uncharted territory. However, there's also something much deeper at play here—what we believe about ourselves.

In his book *The Big Leap: Conquer Your Hidden Fear and Take Life to the Next Level*,[2] Gay Hendricks brings up an excellent point: We get to keep our limitations if we argue for them. What if, instead, we argue for what *is* possible?

Imagine you are asked to give a presentation on a topic you have never taught before, and you have two days to prepare. What immediately comes to mind?

For some people, this scenario seems like a piece of cake. They have already practiced public speaking many times, so it is no problem. For others, it inspires instant panic: *Who am I to be an expert on this topic? What if they find out that I really do not know what I am talking about?*

According to an article published in the *International Journal of Behavioral Science*, an estimated 70 percent of people experience imposter syndrome at some point in their lives.[3] I was speaking with a successful executive in high tech and was surprised to hear that he sometimes feels like an imposter as a leader. Feeling like an imposter is a natural human experience for many of us. Where does this phenomenon come from, after all? Much of it we learn from others in our families and cultures.

Throughout our careers, many of us have had managers who made us feel small, inept or unworthy right when we thought we were showing up as our best selves. This can erode our confidence in who we are as skilled professionals. However, with practice and the right support, I have seen countless people overcome it.

As Sabina's experience illustrates, it is up to us to remind ourselves that we *can* stretch into something new, even when it is uncomfortable—while also allowing ourselves to be okay with failure. Everyone fails at one time or another. Inventor James Dyson had 5,126 failures before inventing his now-best-selling vacuum.[4] Failure is how many of the greatest inventions, stories and advancements in technology, science and humanity have happened. It is also often how teams have breakthroughs, companies are formed and best-selling books are written.

In that spirit, below are four proven, effective keys to overcome imposter syndrome:

Stop Comparing Yourself to Others and Get Clear About What You Want: When you start to look at yourself as a unique individual who has skills and experience to offer, you can stay focused on your own path and what you want to learn and contribute. Comparison will take the wind out of your sails; however, focusing on developing your own passions and skills can propel you forward.

Focus On Your Strengths: Many people lack awareness of some of their biggest strengths, but there are a variety of free resources online to help you assess that, along with what skills you want to develop. You can also work with a private coach to help you reflect on your goals and how the world sees you as a professional. Getting feedback from your colleagues and friends can also be a great way to collect data from those who have seen you in action.

Celebrate Your Wins: If you do not see the value you offer, you probably are not celebrating your wins. Celebrating even the smallest win, such as a customer call that went really well and ended with a happy and smiling customer, can boost your confidence and help you see that you're adding value and making a difference at work.

Find Fun Ways to Celebrate During the Workday: Blast your favorite song. Even better, dance it out. Call a friend and share your great news. Write down what you accomplished and keep a list of your wins.

With some new habits and self-discovery, you can shatter your imposter syndrome. A mentor or coach can also support you in seeing your strengths as you grow, and can catapult you on your path to success in your career. Imposter syndrome is a natural experience in a world that often measures success by specific outputs, and that constantly makes success a moving target. Staying focused on *you* and what you have to offer can be the steady compass that navigates you to your most confident and successful self.

Lastly, know that if you are experiencing imposter syndrome, it is entirely normal and that you are not alone. Just remember: there are tools to overcome these limitations, and every aspirational woman and mother in tech has had to do so throughout their career in their own way.

Keeping this in mind, let us move away from this topic and into the particular: the stories of the tech women blazing forward, and the inspiration behind this book.

PART 2

CELEBRATING MOTHERS IN TECHNOLOGY

FOUR

FINDING JOY AT HOME AND WORK

AMANDA'S STORY

66 "On Sunday nights, I am dying to go back to work. And on Friday afternoons, I am just dying to focus on my kids. I love that, and I think that for the right person, it's just an amazing thing."

—Amanda Berger, Chief Customer Officer, Board Director and Chief Officer at HackerOne

In a world so challenging for women, Amanda's story is a heart-warmer, bringing hope and cheer for women in tech, inspiring them to never give up as they chase their dreams of getting married (or not), having kids and living the life they have always wanted.

Based out of the San Francisco Bay Area, Amanda has worked in business-to-business (B2B)

tech all of her career—which spans more than 20 years—and has fluency in several different protocols such as business intelligence, service security and e-commerce, along with her current operational field of cyber security.

The biggest achievement for Amanda, however, is how she managed to get married, have three kids and maintain an executive position for 10 years. Although it is an outcome many ambitious women have come to expect, according to a recent survey, 91 percent of working moms still expect to pay the "motherhood penalty" upon returning from maternity.[1] Here is Amanda's experience navigating tech motherhood in her own words.

* * *

What was the experience like being a woman in tech who was trying to get pregnant, having a baby and balancing an executive-level job?

I became VP for the first time when I was 29, so I had quite a number of years of executive experience even when I got married at 34, and then it took us a couple of years to have kids—I have twins, who are almost six now. We had a hard time, actually. And then came a 'surprise baby' number three—a daughter, one-and-a-half years old now.

I left the company I was at back then because negotiating maternity leave was not working out. I wanted to take more time off than they were offering me, so I decided to depart as it just seemed like a better solution. I had no idea how hard having twins would be—and about three months into my maternity, I actually did not know how I felt. I had wanted children badly, but I wanted to go right back to work, and then I did not want to work at all. Then I realized, 'Okay, I can not really go back to work.' My life went from being this big global thing to me in my living room. But I was lucky; I ended up getting another job when my twins were four months old.

I went back right after the twins turned six months old, and pretty quickly got back into the rhythm of traveling a lot. It is interesting because it is a very sharp contrast to what the early life of my third child, baby Scarlett, has been because of the pandemic. Looking back, I feel my time with them has been richer, and I have been a more patient mother for it. But starting when they were nine months old, I was gone fairly frequently. Of course, I missed them. I spent the early part of my motherhood bouncing between 'I cannot wait to go to the office' and 'I cannot wait to go home.' And it has been like this up until COVID-19 hit us all.

I actually started the job that I have now at HackerOne after the pandemic.

How did you transition into your role at HackerOne?

I started thinking about what I really wanted to do and I realized that I was not going back to the job I held. I thought maybe I could get into fertility tech as I am quite passionate about supporting women as they go through fertility treatments. I had some connections at places like Kindbody, but I did not have any business-to-consumer (B2C) experience, and even if they offered me a job, it would not have been the caliber of job that I had the leadership experience for. Then I thought maybe I should try to get a non-profit job. But it came down to the fact that it would have been a huge lifestyle adjustment. I had spent all these years in tech, customer success and professional services, and I had these 21 years of knowledge and expertise. I felt strongly that I should not lose them, so I ended up in HackerOne, which is a platform that connects benevolent ethical hackers with companies to perform ethical hacks to ensure the internet is safe. What I really love about HackerOne is that we provide so many opportunities across the world and completely change lives through our technical skills, and that feels good.

What do you wish you'd have done differently that you'd like to pass on to other women?

I wish I would not have fretted over the decision-making process around having children and growing my career in technology—I got very stressed about whether or not I would have kids and how it would happen because I had a hard time. If I could have looked into a crystal ball back then and known that I would be sitting here with three kids, I would not have wasted so much time worrying about having kids while being an executive. I also felt there was a big stigma associated with fertility treatments—I was not very upfront about it myself. But I do wish I had worked in an environment where I could have been more upfront. Going through these processes is painful and emotional, and I think it should be okay to take that time off or have that mental break whenever you need.

Do you feel tech companies are changing in how they are supporting mothers?

I definitely feel maternity leave has improved. In my first company, I was getting nothing but FMLA (Family and Medical Leave Act), but FMLA is not a good thing. By the time I had Scarlett, I ended up getting four months of fully paid maternity leave, and heard some companies were giving six months. So, I do feel that is a big change. Now that I run a big organization and have three young kids, I think the work-life balance I have worked out is wonderful. With the work-from-home culture, sometimes my babies sit on my lap while I attend conference calls. Dads, many of them CEOs, are now having similar experiences due to the pandemic, and it has resulted in more appreciation for working mothers. It is more holistic to include a flexible work schedule, especially for new moms, and I think that I am unapologetic about it. If I have to have my kid on a call, I just have to. If I have to take my kid to the doctor, I have to do it. I am unapologetic about that, and I think that with more women in leadership, there is more opportunity for that kind of change.

For your company, what are you speaking up about? What are you giving mothers in technology that you may not have received yourself in the past?

Maternity leave is one of them. Yes, my company is really understanding about that, but I just think that we have different issues right now with COVID-19. I am really lucky that my children got into an independent school, and they have been going to school throughout most of the pandemic. I have an au pair to support our family as well, but not all parents have this luxury. Companies need to realize that right now, a lot of women do not have any child care, and with teachers conducting virtual classes, it's all the more difficult. My kids do virtual school sometimes, and I have to take responsibility for them. They are sitting there and say 'I need an orange', 'I need to go to the bathroom.' It is really hard. I think there needs to be a heightened awareness that there's no way around nature and that it is really hard for mothers now. I am sure it is hard for fathers, too, but I do think it is harder for the mother in most cases. My company is very understanding about it, but no matter how understanding a company is, it is hard not to feel guilty. That is the other thing I want to say—do not feel guilty. So what if I have to take my youngest to the doctor for an hour and miss a couple of hours in the morning? I think it is important to tell people, 'do not feel guilty.'

Today, as I speak about this, I realize I have been given this opportunity to do something good for the world, and I actually think my daughter, Scarlett, sort of inspired me to do this. At one point, I told myself, 'If I am not going to sit with my three little ones, who I wanted so badly all this time, I need to do something else that's positive. I can not just make money for companies!' My kids ask me what I want to do, and I tell them I make the internet safer—they are too young to get it and ask, 'Can you show me what you made today?' And I have a good laugh. Soon enough, I will."

As a parting thought, what is your advice to mothers in tech thinking of adopting or having a child?

First of all, I would tell them to do it! I think that having a child and being in technology is a wonderful opportunity to demonstrate to your kids how to live your life. One of the greatest things for me today is that my kids see me as somebody who influences the world, changes the world—and it is really important that they do not just see their father that way. Additionally, one of the best things that has happened to me ever since having kids is that I interact with some of the people at work differently because we're all parents. As working parents, I now know that we need to talk about kids not sleeping, teething, being mischievous, as much as we need to talk about business. I think it is okay to be a mother and a technologist, and also an executive. In today's world, those three things go together in a way that our mothers never thought that they did. And that is the thought I hope that aspiring mothers embrace.

* * *

What We Learned from Amanda:

As Amanda's story shows us, things were once much tougher for working mothers, with next to nil recognition of their familial responsibilities (in some cases, resulting in a struggle to even get adequate maternity leave). Like Amanda, Emilia faced a similar struggle:

"I also did not get any maternity leave in 2011 while working at Jobvite. Instead, I had to save up five and half weeks of vacation time so I could spend time with my baby. The excuse was that 'I did not live within 50 miles of the office and the company did not have enough employees to have to offer me any maternity leave.' At the time, I accepted it as it was, but you can be sure I also started looking for another job. Especially knowing I had dedicated five years of my career to this and this was their thank you. It is a real thing that women older than 30 are put on an invisible 'maternity risk list' by their bosses. In countries like Italy, where I lived in my mid-20s, employers flat-out told me they did not want to hire me because I would be having children soon."

The reality is that maternity leave still remains a major void for women in tech, though lately, with employee turnover having burned one too many companies, tech firms are beginning to heed female workers' voices. To combat female staff churn, many companies have started to give longer maternity leaves. Google has extended maternity leave for its women employees to 18 weeks from 12, and also announced that it saw the attrition rate for new mothers fall by 50 percent. Microsoft and Amazon (famed for its intense work culture and lack of sensitivity towards the lives of their employees) also raised their maternity leave period, with both offering 20 weeks of paid leave for birth mothers.

Still, not everyone has been so fortunate during the pandemic. In many ways, the pandemic has only amplified existing biases. Mothers were considered a liability at work even before a virus turned the world upside-down; now, with work, child care and school all happening at home, working mothers are at the risk of paying a hefty motherhood penalty all over again. In fact, multiple post-COVID surveys predict that mothers, especially senior-level women, face distinct challenges at the workplace that could end up prematurely pushing them out of the workforce.[2] Maternity leave aside, lack of flexibility and workplace environments that make women feel distinctly unwelcome are the two main reasons why working moms actually exit companies (not just tech). Contrary to what employers believe, flexibility fosters better employee productivity. Research shows that workers with flexible schedules showed 53 percent higher productivity and accounted for 57 percent better work-life balance.[3]

An interesting complication in these negotiations is how women feel accepting space and understanding from their employers, even if what they are offering should be freely given and accepted without second thought. What Amanda notes in her own story is echoed in *The Wall Street Journal* about how working women can "ditch the guilt" when struggling to find work-life balance.[4] The essay explores how women in leadership roles navigate the corporate ladder and raise families at the same time. Above all, the article highlights how American society is gripped by a "working-mother guilt" syndrome and gives women tips on how to adopt a more unapologetic mindset to combat it.

An important detail to add to Amanda's story is that luck worked in her favor. A former male colleague called her and asked her to work for him while she was on maternity leave and before searching for a way back into her career. When she hesitated, he created a position for her and held it until she was ready to go back to work, when the twins were six months old—an unheard-of scenario in the tech world back then.

Above all, what is truly inspiring is that in and out of tech, Amanda made positive, life-changing decisions after wading through all the bro-isms of the tech world —and it was not tech that gave her the inspiration to do so. Instead, the affirmative changes she made towards work satisfaction, work-life balance and the feel-good factor was because of her kids—her youngest Scarlett, to be precise.

Amanda's inspiring story is distinct, as she was granted a significant degree of support as she returned to the tech industry after becoming a mother. While this may seem unusual or not the norm, this is also encouraging of the fact that there are people who are willing to support and work with working mothers. The support and cooperation she received from her colleagues, as well as the closeness of their bond as parents, shows that even though women still have to overcome a lot of obstacles, there are those who share in their difficulty and do their best to alleviate it.

We have also learned that the greatest challenge to potential mothers can often be their own mindset. As Amanda put it, worrying about the circumstances and the future was one of the major aspects holding her back from realizing her dream of being a mother. Once she pushed that worry away and decided to stick to the path she wanted, she was able to overcome those challenges. While there are always roadblocks, Amanda has taught us that being determined and following your dreams is essential to achieving your goals. It is completely possible to be a working mother and an employee of the tech industry at once, and if you do not take Amanda's word for it, then the statistics of our survey will definitely convince you.

Of the 318 women we interviewed, 120 shared they wanted to leave tech after having a child, while the remaining 198 did not consider the thought or declined to answer. The majority of women did not believe that having a child would impact their career, something that Amanda voiced. More surprisingly, only 15 percent of those 318 women left the industry after becoming mothers. This disproves the notion that women with children cannot continue working within the tech industry.

The most important lesson we have learned from Amanda is not just how valuable mothers are in technology, but how successful they can be as well. It is a story that can change and shape the world around us. As mothers, we have the opportunity to become role models for our children and inspire them to aspire to achieve their dreams. We believe that there is no greater gift a parent can give than to inspire their children to do good while embracing their passions and strengths in the world.

What Tech Leaders Can Learn from Amanda:

While Amanda's willpower and determination

were key to her success, she had multiple people to support her along the way. The most significant of these was her old boss, who not only offered her a position while she was embracing parenthood, but also held the position open for her until she was ready to take up the offer. His act is a rare occurrence in the industry, but tech leaders can learn quite a lot from his generosity. The opportunity he offered Amanda helped her in a time of confusion and showed her a way back towards achieving her dream. In return, she contributed greatly to the company and provided her expertise to ensure that the company soared to greater heights. Though many tech leaders today might never consider doing something similar, we believe that should change. Women should also be offered opportunities in consideration of their circumstances. In the past, tech leaders have only considered short term results, but we believe that they should start looking at the bigger picture. There are thousands of skilled mom-technologists all around the globe who are capable of changing the world. Tech moms have already proven that they can bring prosperity and success to the tech industry, and leaders should start offering them more opportunities so they too may realize these successes. There is no rule that a parent cannot be a successful tech worker, and it is time that the tech sector embraced that idea.

You focus only in tech side
What about the future of
family? - Result after 15/20y

FIVE

CLAIMING YOUR OWN PATH

JACKIE'S STORY

> "Now, I am at a stage where it's about what I want to do—my career, my time, who I want to give back to and what significance I want to make in the world. My children are raised and adults and I am not a helicopter parent trying to live their lives, I am living mine."
>
> —Jackie Rodriguez, Vice President of People and Culture at Givelify

What must it be like to live as a Latina in the United States of America? On top of that, what must it be like to be a Latina in the male-dominated tech world while being a mom, too?

According to a recent study by the Working

Mother Research Institute, climbing the ladder is so hard and slow for multicultural women in senior roles that only one percent of them make it to the top. It is demanding work that involves making sacrifices, jumping over hurdles and overcoming discouragement that can make you question your self-worth.

In spite of this fact, Latina mothers have been quietly disrupting the tech industry and have been influential in changing the very face of technology as we know it, though they have had to overcome a lot to succeed. As Jackie explains, her story is no different.

Unlike some of the other women interviewed, Jackie's career began in human resources, but as the only bilingual person at the hotel that employed her, she was able to leverage her unique skills to learn the ins-and-outs of HR before applying those skills in entirely new contexts.

The average woman in the US earns 80 cents for every dollar that a man earns—but for Latina women, that number typically falls to 53 cents per dollar.[1] Surveys show that of the total computing-related jobs in the US, a mere two percent went to Hispanic women.[2] As Jackie's story shows, she is an exception—although she has faced more than her fair share of discrimination and adversity as a mother of color in a bro-culture to make that so. This is her story, in her own words.

* * *

How did your career trajectory in tech begin?

I grew up in South Texas, on the Texas-Mexico border, and I was raised by a single mom. I graduated from the University of Texas, Austin and was all set for a career in education as a teacher when I stumbled upon a career in human resources. I got a job as a bilingual coordinator at a luxury hotel, and 80 percent of the staff there spoke only Spanish. As a result, I got to learn the whole gamut about HR because I was the one who had to translate everything.

By that point, I had already gotten married. I married young and soon had children, and from there, it just kind of took off in terms of career. I developed a passion for human resources and I ended up being recruited to an ad agency that served blue-chip Fortune 500 companies for Hispanic marketing. After several years of that, I went on to work for Alyka ad agency—a digital one. Those were the early days for tech. I got to see up close so many situations and things. Curiosity also got me to do actual marketing assignments, and so I had clients—which, by the way, is how I met Erin Dovichin, [the Managing Partner at Alaska Venture Fund]. She was our client at that agency.

I handled Latino account planning and slowly grew in my HR career. As I gained confidence, I decided to try a much larger company—I joined TGI Fridays. There, I helped them set up Cinépolis in Mexico to establish a strong US presence. Givelify is my first foray into the high-tech world, and it happened much later in my career—and this, despite being married for 32 years to a tech guy! So, it has been interesting for me to take my core expertise in HR and apply it in a completely different industry. It is exciting, and I am learning and growing.

Have you inspired your children to be interested in tech?

I am now mother to three adult women ranging from 21 to 30 years old, and they are all pursuing careers of their own. My eldest is the head of international sourcing at Topoff Talent—they work the engineering and tech side of sourcing. The middle one is a nurse and was right on the frontline in the pandemic. The youngest one, Julia, is a student. Her interests lie in public relations and planning, and so she is doing integrated studies in communication. When COVID-19 hit us, we at Givelify were working 24 hours a day for 7 days a week, trying to reach out and support. We could not hire because we did not know how the business was going to go or whether it was going to last—nobody was sure about what would happen. Then Julia's internship got canceled and she was home hanging around, so I decided to put her to work. I trained her to answer customer questions, and now she is working in customer care at Givelify.

What did it feel like being a mom in the tech industry?

I remember being in a conference that had Michelle Obama and Maria Shriver on the stage, and they talked about how regardless of your level or where you are at in your career or success, there is always that element of walking into a room and feeling, 'Should I even speak up?' The trick is to think 'Yes, I am making a recorded contribution to whatever is happening right now.' Sometimes the barriers you face are perceived and are those in your mind—so you have to resolve them up there.

Do you think this will all change for future generations?

I think so. If you can intentionally go out there and

tell other mothers, 'Hey, what you are experiencing is normal,' and you have the vulnerability and the transparency to admit, 'I fought that, too. You are not alone and you are not the only one.' If we can go out and say that struggle is normal and that it is common, tell them they can overcome it and just encourage one another to put yourself out there—change will happen.

What about work-life balance? A marriage, kids, career, husband, working—how did you manage everything?

We developed a support system. In our case, it was our love for the church—and we had very close friends who lived similar lifestyles and had similar interests. That helped a lot, particularly when the children were younger. The other thing is, I was okay with where I was in life and at what stage I was at. I could not do what I do today when my children were young because they were my priority then. So maybe I delayed getting into a C-level position or pursuing that when I was at that stage in my life. For me, that was okay. The time will come when you have a career—but the time with your kids is so precious. They are gone before you know it.

As a mentor, what would you say about the support mothers in the technology business seek today?

I would like to see more women, especially mothers in the high-tech virtual world, have a clear vision about the conversations they are having about more safety, more opportunities and the freedom to be vulnerable. I would like them to come out and say, 'I am struggling with this—has anybody had a similar situation and how did you resolve it?'

What advice would you give to women thinking of going into tech and becoming parents?

Do not make assumptions. Have conversations with leaders, with others. Be a catalyst for change. Maybe your CEO never thought about what is on your mind and did not know there was the need for that—so, do not think it will never change. It will, and if you are in an organization that is not open to your needs, maybe it is time to look for one that is. Also, do not be afraid to make the decisions that are right for you. Not everything has to be right now. You do not have to sacrifice everything for work or serve everybody around you so that at the end of the day, you do not have anything left for yourself. It is okay to make choices and leave some for others—just prioritize. Ultimately, it has to be what is healthy for you. If you do not take care of yourself, if you are not in a good place, you cannot give to others what you inherently do not have within you.

Finally, there is always someone else who is up and coming to where you are wanting to be or have been before. I fundamentally believe that you reap what you sow, therefore remember to help other women who have been in less privileged situations.

What are you most proud of in your life or career?

That my mama is proud of me. She is an amazing woman who found herself single at a very young age. I was about two years old, and to support me and my infant brother, she joined the Navy, served the country, brought in income and was able to support us. She made some very bold decisions to do what she had to do for her family, and I have been so inspired by her. She has made sacrifices, worked hard and never let the fact that she comes from a poor little Texas town or her Latina-plus-single-mom status deter her from pursuing a career and giving the best version of herself to others.

This has been a huge catalyst in shaping who I am today.

What We Learned from Jackie:

Jackie's story shows us that mothers in technology must pave their own way forward, as no amount of racism or sexism can deter a mother from being a successful tech worker if she sets her mind on her goal. She has shown us that women must make decisions that are right for them to guarantee their triumphs. As long as they stick to their decisions with true determination, they are bound to find the success they deserve.

Perhaps the most important thing that Jackie has taught us is just how important it is for mothers to support other mothers and women, and to help them move forward in their careers. Not only has Jackie established a precedent for other mothers, but she has also motivated her three daughters to pursue careers in technology that could help change the world. She has also continued to advance her skills and knowledge to help other women to the best of her ability, which is something we should all aspire to achieve.

Jackie has also shown us the importance of appreciating the efforts of mothers, both in and out of tech.

Her heartfelt appreciation of her mother's struggle shows just how important her contribution was to the achievements of her daughter. With more inspiring women at the forefront, there is no doubt that they will lead the way for women establishing their merits in the tech world.

What Tech Leaders Can Learn from Jackie:

Jackie's journey through the tech world offers deep insight into the capabilities and enthusiasm of women towards technology. Tech leaders must realize that these women are deeply passionate and possess the skills necessary to take the sector to new heights. Allowing for more women leaders, especially mothers, to establish themselves and act as role models can offer an incentive for aspiring tech entrepreneurs and increase the influx of great minds into the sector. This can, in turn, result in innovation and new perspectives that can reshape the way we look at things.

Jackie also offers another perspective that tech leaders should consider: there is not enough communication between mothers and tech leaders today. This is why many mothers feel that their needs are not catered to. While it is true that when women speak up, their voices are often ignored, it is important for conversation to occur between tech leaders and mothers. This will at least ensure that tech leaders are aware of what mothers expect of them. Therefore, tech leaders should make an effort to go out of their way to ask about what working mothers in tech require, and to try to make necessary arrangements to ease their journey towards success.

SIX

LISTENING TO YOUR INNER VOICE
PRIYA'S STORY

> "There are loads of companies where you can prioritize both motherhood and your job, so set rules for yourself even before you enter a company."
>
> —Priya Ramachandran, Vice President of Engineering and Operations at BetterCloud

Not all battles are won at the workplace. Some fights are internal and have to be fought in our heads. There comes a time in every working woman's life when despite being shredded apart inside, she has to put on a brave, smiling face, go to the office and get the job done. A time when she has to prioritize some tough choices and make the calls that are best for her.

It's not easy—especially if you are a woman of color

in the US. Top that with being in tech, and you are beset with insecurities about what will happen to your career were you to get married or have kids. But then, some can turn even these experiences into positive ones, and help other women learn and shine—much like Priya, who lives in Foster City, California.

After getting an engineering degree and working as a developer in India, Priya got married when she was 22 ("Back then, the two questions on every campus interview were, 'When are you planning to get married?' and 'Will you quit your job when you get married?'" she jokes). After that, her first major project was for Nestlé in Switzerland, after which she moved to the USA for a job at Keen Consulting. Next was LogiGear, which she calls her most important career shift, and a few years at Intel Security setting up teams in India.

After this came a role at Coupa, though after a few years, she decided to take a break and chart a new path. In consideration of her two boys, who are three and nine years old, Priya paused her career to focus on being a mom—though she continued coaching women around the world in the interim. After returning to work, Priya was able to get her director post back and grow to the level of senior director at Coupa, as well as get on its Empower board for women.

Since then, Priya has left Coupa and is helping an

ex-colleague start her own company, coaching women, working with INSEAD and mentoring their Women in Business group. She has also signed a pro bono contract with Freshworks to work with their engineering team on coaching women in IT and technology, accepted a role as Vice President of Engineering and Operations at BetterCloud and served as an advisor, mentor and angel investor to various other companies.

Priya has an atypical immigrant story in that she came to America with a job in hand, but that does not mean she has not had the same demons to slay as thousands of tech women in the USA.

What is your personal story? How did you get into tech and climb the ladder high enough to be able to help other women, too?

I was born in India, in Bangalore, and I did my engineering degree back there. I lost my dad when I was 15. From then on, it was my younger sister, me and my mother. My mother is one of the strongest women who is always inspiring me consistently. There is this beautiful story that I would love to share. My childhood was quite an easy one, as my parents were financially stable. My mom held a master's degree and my dad was an engineer who was into business. But even with being financially independent, when I turned 18, just because I did not have my dad, a couple of my family members came home one day, telling my mom that I should be married off. I still remember holding my sister's hand, terribly scared that my mom would accede to their wishes. Today, my sister and I talk about it and laugh—she's a CMU grad and is here in the Valley, too. But back then, I remember my family members were trying to convince my mother that since a lot of money needed to be spent on my wedding, spending more on my higher education by sending me to an engineering college was a foolish thing to do.

My mom looked at these relatives and said, 'I thank you for your concern, but I am never going to come and knock on your door asking for any financial help. We decide what is important for our children, and education is the most important thing that I can give to my kids. So please do not ever come here again advising me about what I should do for my children.' Even now, I get goosebumps thinking about it. That was a solid moment in my life that proved to me that my mom would always support us and look out for us.

Though you have had many experiences in

tech, which one has most impacted who you are today?

It was that stint in LogiGear, the team back in Vietnam and all the customers who affected that major change in me and helped me develop into a better person and a good leader. It was the beginning of a journey.

I was the first woman director in LogiGear—and we'd meet a bunch of men to close deals. I remember during one such meeting with a Fortune 500 company, when proposing certain process changes, one experienced senior manager told me, 'You are about 27—do you know I have that many years of experience in this company?' I just laughed and said, 'But still, sir, I am the one standing here and proposing to you what needs to be done.'

The whole world is talking about #MeToo and women's empowerment today, but I still remember that even back as a part of LogiGear, I was going to universities and talking about how we would love to hire, promote and create an environment for women and women of color. I always had a different perspective on hiring. I always thought attitude was bigger than checking the right requirement boxes on the resumé.

I did observe, though, that 90 percent of the time, I was the only woman in her mid-20s with a group of male co-workers. I used to come back and tell my husband, 'Again, it was just me and a bunch of men.' But I was too young to know that this is what 'lacking diversity' actually means.

How did having kids change your attitude towards tech and your own career?

My first child was born in 2010 when I was traveling a lot. I was working towards a senior leadership position at Intel. I had the best leader anyone can get—Jamie Tischart. A leader with empathy. He used to be my customer when I was in LogiGear, setting up Vietnam practices. He was my boss at McAfee Intel Security. I just spoke to him and told him I wanted to take a little break and then join a job with no travel. He asked me, 'Priya, are you sure? I am confident you can grow any time you want to, but Intel is amazing for women's empowerment.' They had already put me on that path with training. But I was sure. That's when I applied to Coupa.

I took a role that was one level down, but I did not care about that—I wanted peace. Mainly because I had had about four miscarriages after my first child by then. From 2010 to 2015, it was terrible. I wanted to grow and to coach women, but I seriously wanted to expand my family, too. I spoke to Vilas Madapurmath and JP

Krishnamoorthy of Coupa and told them I did not want to travel. The workplace was five minutes from my home, and I was able to conceive and have my baby. It was good.

I spent a lot of time mentoring women, and went ahead and did my MBA from INSEAD. So, with kids who were two and eight, I managed an MBA and a full-time job where we used to work 12-14 hours.

There have been people who ask me, 'Why are you getting your MBA?' My husband did his MBA in 2013 from UCLA Anderson, and not a single person asked him, 'Who takes care of your daughter?' But I have been asked again and again why I am investing in an MBA instead of spending time with my children. No one knows that I was still doing everything that I needed to be doing with my kids, even when I was in France (INSEAD). But my husband was supportive—he said I must pursue my dreams. The same thing happens at work also. Even executive leaders have asked me, 'Why are you working so hard?' I want to ask, 'Would you ask this question to a guy?'

Why do many women talk about leaving technology when they become mothers, **especially women of different cultures**, **whereas some choose to stay on? Are the**

two mutually exclusive? What makes the women want to leave?

It is a combination of many factors, in lesser and larger degrees. Many people do not even focus on the whole subject of how women need their voices to be heard. So though Freshworks wanted to pay me, I said, 'No, not at the moment.' I wanted to work with them because of their intent and approach to the whole issue of women in tech. And I thought, 'I want more companies to become that kind—helping and promoting women.' Even today, I spent one hour talking with one of the ex-Coupaians from India whom I have been mentoring. It is just so important for us to stand with each other.

I was fortunate. I was able to do my MBA from INSEAD when I was in Coupa, and by the time I got into a job after that, I intrinsically knew that I just wanted to give back what I had received. I got to spend time helping out women and supporting them, because that was also the best way for me to learn something in return.

How can tech companies and leaders create better working environments for women?

At Freshworks, the CEO, Girish Mathrubootham,

was so prompt. The very next minute after I mentioned to him that I intended to spend time mentoring women in engineering in his company, he initiated the program's implementation. That level of interest is what is going to enable an inclusive workplace. It is not about diversity, but rather it is about *inclusivity*. If every single CEO makes that kind of an impact from the inclusion perspective, practicing rather than just preaching, it will be great for all the mothers out there.

Workplace gender discrimination is indeed driving away women from tech in the US[1], right?

Yes, and I do not want the women I coach and mentor to feel this. I tell them, 'Doing what is good for you, as long as you do not hurt or backstab others, is the best thing to do.' I took a 42 percent pay cut when I moved from Intel to Coupa because I wanted to focus on my family, but no one thinks about it. In May of 2021, when my mom was down with bilateral pneumonia, we persuaded her to come and live with us in the US. At that time, I converted my FMLA into a break because my son was three, my daughter was having her summer holidays and I had to support my mom and children—but no one bothers to talk about that. When we give our best, people ask us why and judge us for doing that. I tell women to respect and not judge what everyone is doing. It is good humanity and good leadership, too.

It has also given me this idea of creating a product on DEI (Diversity, Equity and Inclusion). The greatest news is that with the vaccines becoming available, my mom is set, my daughter is back in school and my son started preschool. I finally have more time at hand now!

What is your advice to women who are struggling to stay in tech and start a family — perhaps those that are going through fertility treatments, having miscarriages?

It depends on the individual at that moment in time. Talk to yourself about what is important to you. There are loads of companies where you can prioritize both motherhood and a job, so set rules for yourself before you even enter a company. Get a coach, a mentor, have open conversations with them, run your ideas past them, get different perspectives and see who you relate to the most. If you take up what you truly want to do, you will grow. So, talk to your manager. Be clear. If your leader cannot understand that, it might not be the right place for you. So, look for a new job, but never doubt yourself. As I keep saying, what is needed is work-life separation and not integration—this will help employees have good mental and physical health. And that is important for society, the company and the family at multiple micro-levels.

What We Learned from Priya:

Priya's story is perhaps one of the most inspiring within this book and offers us clear insight into how a mother can establish her place within the tech sector. Even with adversity that Priya has faced, she has always stood her ground and fought back through her words and actions. She has proven time and again that she is equally capable as any man, and motherhood has never held her back from success.

More importantly, Priya shows us how important it is to work out a balance between work and life. During work hours, Priya is a dedicated professional who gives her attention to her job, and during off-hours, she ensures that she dedicates her time to her family and her children. This balance has proven key to her success as both a tech worker and a mother.

Priya also shows us just how important it is not to buckle under societal pressure. Regardless of all of the stereotypical notions she had to face within the work-place, she never gave in to any of the pressure and always put her career and goals first. She challenged the notion that mothers have to sit at home and tend to their children, and proved how wrong it was by successfully advancing her career while still giving due attention to her children. The concept of good humanity that Priya has so heavily emphasized is also relevant here, as we must ensure that we allow mothers to construct and follow their own paths instead of forcing stereotypes upon them. A mother is perfectly capable of raising up her children and focusing on her career at once, and Priya is a prime example of this.

Priya has also shown us the importance of mentorship and how mothers and women who have experienced the tech industry must provide guidance to their successors and protect them from unwanted pressure. The key takeaway is that women must help other women, and mothers must help other mothers build up their careers, as nobody can be better aware of the common challenges. Individually, a mother might feel discouraged and shy away from her professional aspirations, but collectively, when helping one another, mothers and women will be able to thrive.

What Tech Leaders Can Learn from Priya:

Inclusivity is one of the most important lessons from Priya's story. As she stated, it is about inclusivity rather than diversity. Tech leaders should not only focus on expanding their teams to incorporate more women and mothers, but they should also focus on establishing a framework that allows these women to comfortably and easily adjust within the industry.

According to Sheryl Sandberg's book *Lean In*, 43 percent of women with kids ultimately leave their jobs.[2] Still, this can be understood by considering that not many women are fortunate enough to be in a position to even help themselves, let alone help others. Viewed thoughtfully, this can be framed as an inclusivity issue—and solving it requires a certain amount of interest percolating down from the level of management. Hiring women and mothers in tech is a start, but what our leaders must realize is that they often fail to give these employees a voice. They fail to give due consideration to their needs and requirements. This should be their primary concern to address if they wish for a more diverse community within their workplace.

Corporate leaders can be reminded about the impact of peoples' actions. As Priya stated, it is very easy to talk about respecting the needs of mothers and empowering them, but that is only the first step, and most leaders stop there. To truly go all the way when it comes to women's empowerment, tech leaders should act on these claims and delegate power to female employees as well. They must allow women to mentor other women and give them the platform to express their concerns and needs.

Respected tech leaders remember that gender and parental status are irrelevant when it comes to producing quality outputs. Even if a mother has family commitments, she can still be committed to her career with the same level of competency as her male counterparts and can contribute equally to the company's success. With a positive mindset, agreed-upon performance indicators and determination, women can set themselves up for success.

SEVEN

EMBRACING MENTORSHIP AND SPONSORSHIP

JENNIE'S STORY

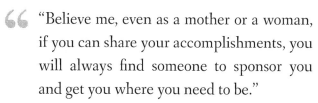 "Believe me, even as a mother or a woman, if you can share your accomplishments, you will always find someone to sponsor you and get you where you need to be."

—Jennie Ibrahim, Software Engineer at Google

In her 12 years working at Google as a software engineer, Jennie Ibrahim has seen changes in diversity and culture that are still evolving to this day. While Jennie spoke favorably of her overall career fulfillment at Google, the media has shown us a different angle. In fact, the viral memo by an ex-Googler titled, "I'm Not Returning to Google After Maternity Leave, and Here is Why," is just one of the many stories that have

circled the web.[1] One thing is certain: the environment, even within one of the largest tech companies in
the world, has shown to be lacking consistent support
of women, and particularly mothers. The bro culture
and male dominance still persist, although it has been
largely replaced with more inclusive policies that have
allowed women to establish a strong presence. During
her 12 years, Jennie has struggled against the odds to
establish her worth at the tech giant.

Jennie has successfully juggled raising three young
children and maintaining an impressive career trajec-
tory for over a decade now. During our interview with
Jennie, she told us that while she works at Google, she
does not claim to represent Google in her interview
responses. For her, individuality is much more important, and she is representing herself—her own actions,
decisions and efforts that she has made to grow to
where she is today.

What challenges, obstacles and barriers have you faced in your professional journey and how have you overcome them?

I would categorize them as two things: One is
having to prove myself over and over again, even when
I know what I am doing. Even now, I feel like there are
times where people that I am leading will ask some-

body else for advice after they asked for my advice. It is frustrating. A man with 20 years of relevant experience would not be made to prove themselves over and over again. He would be given the benefit of the doubt based on his expertise and recognized by virtue of the role to which he was currently appointed.

The second challenge is that there are always changing expectations of my working style. There are some who perceive me as very harsh, or they get the impression that I am not interested in working with them, while others see me as a great partner with whom to work. I am often perceived as very nice, and am told that I should be more aggressive. Alternatively, if I am firm, I am told that I am too aggressive, and that I need to be nicer. The feedback changes based on where I am and with whom I am speaking. It is an internal struggle to figure out which side is lacking and which needs more effort. That balance of being liked and being capable, that is the key, and maintaining that equilibrium has perhaps been my greatest challenge.

It sounds like your working style varies from time to time. Do you feel that you had to alter your approach based on your co-workers' perceptions?

It depends on where the people are coming from.

Who knows what has happened in their past or how they grew up? I'm uncertain about how others are perceiving the situation and what expectations they have of me in the workplace. I can say that when people get to know me, most misconceptions disappear. But for some people, they still persist, unfortunately. For example, in the same role, I've been told 'You're very good technically, now work on your leadership skills.' And then soon thereafter, the same person will say, 'Well, you are a great leader, now work on your technical skills.' I feel like the goal post changes, often without reason. In those situations, you have to realize that you cannot please everybody.

Have you had to change the way you present yourself to be heard or taken seriously?

I took a Cara Hale Alter class,[2] and the way that she talks about how we hold ourselves, the way that we speak and the way that we present ourselves—those are the different ways that people perceive us. And whether we like it or not, they are perceiving us in a certain way. So, we have to have optimal posture, carry ourselves with executive presence and not tilt our heads. We have to enable the opportunity for people to perceive us the way that we want to be perceived. Another thing that I learned over the years is to use data instead of talking about what I think or feel. In the past, on peer feedback, I would say that this person 'did a great job' or 'they are hardworking'. Flowery language does not mean much to a manager. But when you are using data, nobody can question that. Always make it a rule to use statistics so you can back up your claims about your achievements and/or those of your teammates.

Are there other ways that you have had to change and adapt to a work setting, particularly as a mom?

Outside of women-focused Employee Resource Groups (ERGs), where I talk about my personal life and my role as a mother, I keep conversations with other women focused on work-related topics. I am a very social person, so naturally, I get to know people from other teams and other parts of the business. However, instead of just focusing on family-related topics, I try to focus on work and learn more about my colleagues' work.

For someone with your experience, how do you view work-life balance and how do you balance the many roles you have as a leader,

an individual contributor, a mother and a wife?

It is interesting. I asked one of my kids a few weeks ago if he minds that I work and he said that he does not, and he believes that I should be working and doing what I am passionate about. It was refreshing to hear that because I believe they understand what I do because I share it with them. I have them code with me. I make them a part of what I do. And I make them apps, bringing my work to them. My children understand more of what I do and why I like it, and why they might want to like it too, in the future. And even if I am making an app for them, I do not just go off and make it on my own. I try to involve them. Even from the youngest ages, when they can type the letter 'Y,' they can change a color to yellow. I involve them in things related to my work and it becomes a family hobby for all of us. It allows me to focus on work since my children know what I am doing, and they can respect my efforts and let me work when I need to. At the same time, they get a significant portion of my attention as well. Integrating work life into my children's upbringing might sound weird to most, but I believe it is my way of finding that necessary balance. I love having my kids enjoy what I enjoy.

Just as you have been mentoring your children from the very start, did you ever have someone who gave you advice or coach you to help you become the person you are today?

I had people outside of work that mentored me, but having mentors was not the thing that moved the needle forward. For me, it was having sponsors. The difference between the two is very important: Mentors can empathize with you, sympathize with you, and they can give you advice that can help you in the field. Sponsors, on the other hand, advocate for you and help you move up. They are in the room where others decide your future when you are not in the room. Those sponsors advocate for you and let others know about your accomplishments.

I have recently had some amazing sponsors that were career-changing, but one of the most important things to remember is that nobody will speak up for you if they are unaware of your achievements or current initiatives in flight. Make your contributions known. Broadcast your work and efforts to sponsors, and potential sponsors, so everyone knows of the impact you are making. Believe me, even as a mother or a woman, if you can share your accomplishments, you will always find someone to sponsor you and get you where you need to be.

How can women speak up and stand out from the crowd in such a way that there is still a life-work balance and where they can still be an involved mother to their children as well?

I do not believe that working at the next level means that you have to put in more hours, or else the CEO would never sleep. Standing out does not mean that you have to do more, it means that you need to think strategically about what would bring the most value to the team and deliver well on the highest priority items. Delivering incredible work is important, but having a manager who understands your professional value and developmental milestones is essential. If your manager fails to communicate your successful deliverables to her or his manager well, you need to talk to your skip manager about your accomplishments. I never knew that you could meet with your manager's manager or that such a motion was appropriate. Taking this skip-level meeting approach can be done and it is necessary, especially if your direct manager does not see your efforts.

How do you think the industry can support women and allow them to grow in business and technology?

Managers have a big role. And I think there were times in my career when managers would quiet a room to give me the chance to speak up. Or instead of answering questions from their end, they referred other people to me, and that made me feel like I was important and was being given due respect for my experience. Things like that make a difference. It is that support from those managers that has made it possible for women to stay longer in the workforce. These women face hard situations where people make them prove themselves over and over or keep changing the goals, almost every day. If they know that their manager is behind them, they can do whatever they aim to achieve. They can thrive. That is the vital support that I have received. I am so glad I had people like that to support me.

What We Learned from Jennie:

Jennie's story teaches us many things, but the most valuable lesson is to speak up about your own contributions and what you are bringing to the table. In their book, *How Women Rise: Break the 12 Habits Holding*

You Back From Your Next Raise, Promotion or Job, Sally Helgesen and Marshall Goldsmith argue the same thing and squash a common misconception: "Speaking up about what you contribute and detailing why you are qualified does not make you self-centered or self-serving. It sends a signal that you are ready to rise."

Another key learning from Jennie is to confide in industry leaders and have individuals who are willing to vouch for you. "Sponsors" as Jennie calls them, are crucial to the development of your career, and you must ensure that they are willing to stand up for you and defend you even when you're not in the room. According to a study published by *Harvard Business Review,* women tend to be over-mentored and under-sponsored.[3] Having a mentor increased the likelihood of promotion two years later for men, but it had no effect on promotion for women. On the other hand, the benefits of sponsorships are clear and especially impactful for women. In a separate *Harvard Business Review* study, it was shown that "a sponsor confers a statistical career benefit of anywhere from 22 percent to 30 percent, depending on what's being requested (assignment or pay raise) and who's doing the asking (men or women)."[4] As Jennie's experience proves as well, an effective support network can be key in growing a career in tech.

Though there are always changing expectations within the tech field, it is very necessary to adapt to circumstances and present yourself in a manner that is suitable for every situation—this is particularly true when it comes to women and mothers, as the way they are perceived can heavily affect how other individuals interact with them. Unfortunately, Jennie's observations about the additional pressures put on women to behave a certain way are backed up by other research. A 2021 report by TrustRadius showed that 78 percent of women in tech feel they have to work harder than their male co-workers to prove their worth.[5] The study also found that women in tech are four times more likely than men to see gender bias as an obstacle to promotion, and 39 percent of women saw gender bias as a barrier to promotion in 2021. Though this is a difficult state of affairs to be in, being able to understand the requirements of any given situation is key to being taken seriously.

One of the most unique aspects of Jennie's story is her use of data in the workplace, which is a great tactic for aspiring women to use. Justifying one's progress through irrefutable data is a great way to establish worth and clearly define progress and achievements. It is impossible for anyone to deny facts and figures, so using these to your advantage to gain footing in your career is crucial. Facts can be a great defense, so always deliver your best and collect the numbers to back it up.

Lastly, Jennie teaches us that integrating one's work life with personal life can be a tricky business, but if it is done right, it can definitely ease the burden and allow children to develop their own interest in their mother's work. It can inspire children to pursue tech later in their life, and more importantly, it can allow kids to understand the struggles of their mothers. This understanding can go a long way in setting up expectations and requirements. Integrating work within the household as a hobby can allow mothers to give due attention to their children without sacrificing their career prospects.

What Tech Leaders Can Learn from Jennie:

Jennie's experience shows us just how important it is to have sponsors, which can also be instructional for tech leaders. Defending their female employees and standing up for them in the workplace can allow these women and mothers to integrate seamlessly into the company and commit their time and energy towards innovation. It can significantly improve their career prospects and allow them to contribute on a greater scale, both to the company's success and to the inclusion of other women in the workplace. Therefore, tech leaders should give women the platform to speak up, and even more importantly, to defend them when they speak up so that their contributions and input are valued. Tech leaders will realize that the benefit does not just extend to women, but it also reflects on the company itself. An employee who feels that they are an essential part of the company will commit to the company's success even more than their personal growth.

RISING ABOVE YOUR CHALLENGES

ANGELA'S STORY

> 66 "I knew that there was no one who could advocate for me, so I had to do it. It took a tremendous amount of courage, but I knew that I could not get up the next morning and look at myself in the mirror if I did not stand up for what was right."
>
> —Angela Nichols, Advisory Services Consultant at Mavenlink

Our next story comes from a 30-year veteran in the tech industry. Angela Nichols, similar to Jennie, has three children and credits her successful tenure and fulfilling motherhood to being her own advocate.

Even before motherhood, Angela experienced many uncomfortable situations as a woman in tech. One such story of adversity dealt with the type of harassment called "workforce retaliation." In states like California, this behavior is illegal and is defined by the *Society for Human Resource Management* (SHRM)[1] as "a form of unlawful discrimination that occurs when an employer, employment agency or labor organization takes adverse action against an employee, applicant or other covered individual because he or she engaged in a protected activity, including filing a charge of discrimination with a fair employment practices agency or participating in an investigation of alleged workplace misconduct." Not only is retaliation the most common type of harassment in the US,[2] but it has also made recent headlines, with Silicon Valley tech giant Google being sued by a woman who alleged that she received a poor performance review after disclosing her pregnancy.[3]

Aside from a three-year hiatus to stay at home with her children, Angela worked consistently through her children's upbringing and eventually became her family's sole breadwinner so her husband could devote more time to being a full-time parent. When Angela's youngest child was a year old, she returned to the workforce full-time.

Many of Angela's stories are troubling, and other than a story about being put on probation for behavior she believes was in accordance with company policy, she opted to keep the other difficult moments private.

Her story is just as inspiring as Jennie's, as she shows us how to stand up for ourselves and to create room for mothers in this complex place we call tech, even in the hardest of circumstances.

is it not hard work
erasers

* * *

What was the beginning of your career in tech like?

When I started in the workforce [in a tech division within La-Z-Boy] in the 1980s, it was still the case that almost all industries outside of nursing and teaching were fairly male-dominated. It was a real boys network where 'stuff' was happening to me all the time. I remember one of my big awakenings came when I learned that the working world was not going to be exactly what I thought it was going to be. I expected that my career was always going to be an uphill battle. I had to have a lot of persistence. I was going to have to be my own advocate.

What kinds of adversity did you have to face early in your career?

In my very first management position, I was pulled aside by a senior leader of the company early in my term with the firm. This leader warned me about my direct supervisor and asked that I inform him of any misconduct with my direct boss. I was put off by that comment, and then I learned that my predecessor had been a woman who was fired for 'insubordination.' It was a huge red flag. I felt something was not right.

My direct boss was eventually fired and I reluctantly accepted an interim leadership role, but even though I was doing well, the company brought in another man to help run the department. This guy was intimidated by me from day one. There was always conflict between us. He slowly started taking responsibilities away from me and either taking them upon himself or pushing them out to other members of the department.

At one point, I went in to talk to him and I said, 'Listen, you know, I am getting a little bored here because you take away all the challenges in my position. I would like to talk about what we can do differently.' He did nothing to respond to that request. I waited several weeks. I went back and I said, 'Can we talk about this? I brought this up several weeks ago and there has not been any progress.' He said, 'Well, I am really not prepared to talk about this with you.' Three days later, I was scheduled into an after-hours meeting when all other employees had left for the day. My manager was there with a Human Resources representative. I was told that they were 'ready to accept my resignation' and that the grounds for dismissal was 'insubordination.' I was terrified, but something in me said, 'I am not going to let these guys roll over me.'

I knew that there was no one who could advocate for me, so I had to do it. It took a tremendous amount of courage, but I knew that I could not get up the next morning and look at myself in the mirror if I did not stand up for what was right. If that meant that I lost my job and got fired, then so be it. I think it was just pure stubbornness that got me through those times. I think it is necessary, too, because these things unfortunately happen so frequently. I have accepted these situations the way they are. However, I still expect to be the driver behind my own career at all times.

How did you brave through all these challenges?

The challenges poised me with the courage and drive to persist when being a working mom got tough, especially when returning to the workforce. I never felt like there was any other option. For example, during my first years as a working mom, I was required to get to work early in the morning. On my lunch break, I picked up my son and raced to take him to preschool. Then I returned to work for four more hours and then picked him up from preschool, took him home, did the 'nighttime routine' (AKA dinner, bath, teeth brushing, reading a book and tucking in bed) and then returned to work for several more hours, only to do it again the next day. Wash. Rinse. Repeat. I did not realize the gyrations that would be required for me to have a successful career and to be a mom.

* * *

What We Learned from Angela:

Like Jennie, Angela also shows us the importance of standing up for ourselves. The most inspirational part of Angela's story was how she never backed down and always took a stand for her passion and her beliefs, and the way she opposed harassment is nothing short of refreshing within a field that, as Angela puts it, is "a real boys network."

For her part, Emilia also experienced similar retaliation as a Vice President for speaking up about unethical practices at one of her previous companies. She was dismissed soon after the incident occurred without cause except to be told, "we need some fresh blood." Along with their stories, research has shown that 75 percent of employees who speak out against workplace mistreatment face some form of retaliation.[4]

Angela also shows us that it is necessary to stand up for yourself, irrespective of the cost. She was willing to fight for herself and establish her own self-worth, even at the expense of her job. Many individuals might be daunted by the prospect of giving up their job just to take a stand, but as Angela shows, having a strong sense of self-worth opens many more doors than working a job where your efforts are underappreciated.

Another eye-opening teaching is Angela's realization "that the working world was not going to be exactly what I thought it was going to be." It is crucial for mothers and women aspiring to become a part of the tech world to properly research the working conditions and the state of the tech industry before entering. Our intention in saying this is not to make mothers feel afraid of the challenges that they might face, but rather to make them realize that they must be prepared to tackle the challenges that will come their way. A proactive approach is always more effective than a reactive approach. Rather than reacting to circumstances and struggling to tackle problems, mothers should be aware of all the options available to them and to have a clear response prepared for any difficulty that comes their way.

As Angela explained in the interview, she recalls that juggling parenthood coupled with a career was more difficult and time consuming than she was expecting. Studies show that the same struggles persist for working moms today. In a 2018 study by Market Researchers OnePoll, under Welch's nutrition programs, where 2,000 working moms were surveyed, it was revealed that moms worked an average of 98 hours a week between their corporate jobs and their family commitments.

These kinds of situations can take a toll on mothers who are not prepared for them or who do not have enough support. Fortunately, there are dozens upon dozens of articles and blog posts offering advice and free downloadable templates for establishing morning and evening routines for working moms that turn the chaos into calm (we also offer a list of resources for you to leverage in the appendix). Despite one's best-laid plans (remember your birthing plan and how quickly it went out the door?), the mad dash to get the kids to school and sports practices all while honoring your adulthood responsibilities can still make those plans irrelevant at times.

Still, the good news is that as Angela shows, it is possible to forge one's own path as a working mom— and research shows that hard work like hers is paying off. In a study spanning 16 years conducted by the US Census Bureau and summarized in the book *Changing Rhythms of American Family Life,*[5] working mothers today are not only spending just as much time with their children as mothers did 40 years ago, but in many cases, they are spending more time.

In all, Angela is the perfect example of finding the strength necessary to overcome any rough patches. She used the law to her advantage and stood up to the men who were treating her unjustly, proving that she was ready to stand up for herself and establish her worth.

What Leaders in Tech Can Learn from Angela:

Women and mothers in tech are not aiming to replace men or bring them down. They are simply trying to establish their place in an industry that has traditionally been male-dominated, and remains so. Instead of pushing women away from tech and putting them in a corner, leaders can offer them opportunities and paths for career growth. Gender, disability and motherhood should not restrict any woman's advancement within the tech industry or any industry for that matter, as women and mothers have proven themselves to be equally capable. In fact, studies show that they are even more capable than their male counterparts, as mothers have made the same strides in tech while actively raising children and fulfilling their maternal responsibilities. The collective growth of the industry must allow these women and mothers the freedom to prove themselves and show the world that they can change the industry even as they raise children. The industry should forgo its traditional male domination and instead create a space where every individual is free to establish their own worth and receive a position and status that accurately reflects their contributions.

According to the US Department of Labor, 70 percent of moms with children under 18 participate in the workforce.[6] This statistic shows that working mothers are passionate about their careers *and* their maternal responsibilities—and it indicates that mothers are truly exceptional when it comes to balancing their work and personal lives. Given the right opportunities, they can excel without a doubt.

PART 3
OUR RESEARCH FINDINGS

NINE

WHY MOTHERS BELONG IN TECH
THE SURVEY RESULTS

We take great pride in the survey we conducted in the spring of 2021, which, to our knowledge, was the first of its kind. Never before has the connection between motherhood and technology been discussed to such a great degree, and our sample size of 318 mothers, answering 12 questions, taking into account all the various perspectives and challenges that mothers have faced during their careers in tech. For the purpose of clarity, we have divided the results into two parts, and we will be incorporating all the key elements of the first section within this chapter. Let us explore the many perspectives of mothers in the tech industry.

Why did you want to join the tech industry?

A third of the women we interviewed focused on a

career in technology from the beginning of their adult life. They fell in love with opportunities befitting their skills and wanted to pursue excellence in STEM fields. The fast pace of the industry was exciting to them, and with or without children, they were committed to combining a career in leadership with raising a young family.

"I have always loved tech," one participant shared. "I did not get an opportunity to major in it in school, but when I knew I could join the tech world by becoming a CSM, I was more than happy to do it. Tech revolutionizes the world." Among our respondents, start-up success in Silicon Valley and other tech hotspots around the world was seen as an attractive opportunity to get away from traditional jobs and try something new with the Steve Jobses and Sheryl Sandbergs of the world.

As we have already learned, women face a great disadvantage in terms of career growth in the tech industry. By pursuing careers in that field, are women clinging to false hope or simply following their passion and believing in their skills? The results show that it is the latter. "I have always been passionate about learning tech and wanted to make my own stand," one survey participant said. "I wanted to identify where I could make some differences and contribute."

Surprisingly enough, 20 percent of the mothers

surveyed found themselves in a tech career by chance or "by accident," having never intended to join the field. "I fell into it and then stayed," one of the participants said. "Technology is not going away, and there are so many different, exciting verticals and industries to choose from." What is even more surprising is that some mothers were actually drawn to their workplace and fell in love with their companies despite all the challenges they faced in a "bro culture." While the 1980s may have started a tech culture that turned many women off (popularly represented by movies like 1984's *Revenge of the Nerds*), it did not completely lock them out. Among companies who included women in leadership positions since their inception, Google stands out by awarding early platforms to women like Susan Wojcicki, Sheryl Sandberg and Marissa Mayer, all of whom are now household names. Even when it was difficult, these women persisted.

From our surveys, we can draw many conclusions about how women broke into "the boys' club," but one major theme is that there are tech companies that have tried to be more inclusive of women and give them a greater voice. For this, we are optimistic; progress has been made.

Where there are women who have landed jobs in technology and loved their experience, there are, however, also many who made a career pivot only to be appalled by the practices they witnessed. Of the women we interviewed, 59 percent shared that they wanted to leave tech at some point during their career due to problematic work conditions and an unsupportive environment. From all the information we gathered, a few key questions arise:

For women who faced adversity in tech, why did they stay?

Many of these women were not drawn by the workplace environment but rather the goals and purpose of the tech industry. Ten percent of mothers we interviewed shared that they found the ability to innovate and change the world inspiring, which shows that they were pursuing tech based on their passion. In a setting where many women were subject to discriminatory practices, these respondents dug their heels in the sand and their tenacity drove the pursuit of their dreams.

Why did women want to leave tech after having or adopting children?

One mother we surveyed put it simply: "The pace is fast and the commitments are demanding." This answer might easily sum up the entire sector and why it is viewed as such a challenge for mothers in the industry. Unmanageable hours and expectations have long plagued the tech business vertical, and many mothers who participated in our study admitted to feeling the same way.

Ninety-three mothers, nearly one-third of the survey participants, shared that raising a child while working in such a hectic, high-speed environment was incredibly tough and sometimes downright impossible. These women reported feeling burnt out as they were unable to find a balance between work and life; as a result, both aspects suffered greatly. Seventeen percent of mothers shared that they were not given due appreciation for their efforts, and found it difficult to expand their career after a while. Several divulged that it felt impossible to continue their careers while surrounded by peers who lacked understanding of the commitments required of parenthood.

Our survey also highlighted the importance of schedule flexibility for mothers working in technology. Without an opportunity to leave midday to watch a child's school play or afternoon soccer game, and then catch up on work outputs after the kids were in bed, it was nearly impossible to be both a mom and a member of the tech workforce. The demands of their work resulted in an inability to meet the responsibilities to their children. In fact, the 16 percent of mothers in our survey that decided to leave tech mostly did so to devote more time to their children. The concept of "work-life balance" was a misnomer—there was no balance at all. Herein lies the problem: for these women, the time required to complete the work and the rewards for doing the work were simply not worth it when compared to the value of spending time with their children.

Discriminatory practices have also pushed women out of tech: over 50 percent of the survey participants stated that they had been subjected to harassment and sexism. To them, not being appreciated for their services while surviving in a place where they were ostracized or perceived differently was sufficient reason to leave. Even the most passionate of individuals surveyed reported having second thoughts about staying in tech.

A majority of the surveyed mothers within tech wanted to leave primarily because of the environment and work culture, as opposed to leaving to explore other opportunities. In fact, out of almost 160 women (just over 50 percent of respondents) who wanted to leave, only two were motivated by external factors, and only 20 were considering another job. As this data shows, the truth is that harshness and discrimination towards mothers within the tech industry pushed them into wanting to leave, and this is deeply concerning. It means that individuals with passion and potential are being driven away forcefully, thereby depriving the industry of unique and essential talent.

Did these women and mothers leave tech permanently?

Of our respondents, 16 percent said that they no longer found it feasible to continue working in tech and left their jobs. While this is not a majority, it is still a sizable portion of the participants—and to extrapolate, one-sixth of the total tech workforce exiting the sector would be a massive blow to the industry. For perspective, for the approximately six million mothers working in tech, we can estimate that nearly a million end up leaving for the reasons outlined above. The benefits and advantages that these women bring to the industry are numerous, so a loss of this magnitude speaks volumes about the loss of potential that the sector suffers primarily due to sexist practices and lack of consideration for the necessities of parenthood. For every prolific mother with major achievements under her belt, it is important to also acknowledge just how many equally capable mothers never end up realizing equivalent professional achievements or dreams, and how this negatively impacts the tech industry as a whole. This is a failure of tech leaders, many of whom seemingly consider mothers working in tech as a liability. However, it is the contrary that is true. When granted equal opportunities and considerations, working mothers can be some of tech companies' best assets.

Despite all of the challenges and reasons to leave noted above, many mothers are still committed to their passion. While 38 percent of survey participants did consider leaving tech, only 16 percent, less than half, actually went through with it. The rest continued to persevere in pursuit of their dreams. These women are an inspiration to us because despite the odds, they were unwilling to give up on their calling. Even the ones that ultimately left tech are an inspiration, because they put their own well-being and motherhood first and refused to accept practices and policies that contradicted their values.

Were these women ever harassed at work by a colleague or boss?

One would expect that in a corporate environment where professionalism is of the utmost importance, women, and mothers in particular, would be safe from unwanted advances or sexist remarks. Unfortunately, that is far from the truth. In fact, out of the 318 women's voices that we captured in our survey, a staggering 54 percent—169 women—shared that they faced harassment at the workplace. Every company and leader of tech needs to take notice of this information! Our survey results, compounded with the hundreds of workplace harassment reports over multiple decades from around the globe, is evidence of the need for change. When women have shared that even their bosses and top management have been involved in the harassment, the fear of retaliation, even job loss, is real. Until there is change within the upper levels of management, it is highly likely that the climate of many major tech companies will remain hostile to women. This survey finding shows that there needs to be a change within the industry from top to bottom so that women may find the comfort they require to pursue their passion.

For women who left tech, did they return after a short break?

There were several women in our survey who made the decision to leave tech. These mothers took breaks to give more time to their families and daily lives, or took time off due to unpleasant experiences. Out of these, 76 percent eventually returned. While this is a large majority and shows that these mothers' passions were greater than the challenges they faced, a significant portion of the women did not return. Twenty-four percent of mothers wanted to continue working in tech, but left the industry entirely. The faults lie at the most foundational level, and until we address these cultural problems from the root, the plants will continue to wither. Mothers will continue to leave tech and will not come back, and that is not only a loss for them. It is a loss for the industry.

TEN

EXPERIENCES TO INSPIRE YOU

THE SURVEY RESULTS

The second half of our survey offers some unique insight into the mothers who make up the tech workforce. To continue from the first part, we will first discuss how the women we interviewed responded in regards to their careers before discussing how many diverse individuals are involved in tech. We will start off with the first question we asked, which offers deep insight into their feelings about discrimination and lack of support for mothers within the tech world.

When you returned to tech post-baby/post-break, did your employer support you in your re-integration?

A majority of mothers returning to tech did not

feel supported as they reintegrated into the workplace. Out of the 318 women surveyed, only 70 percent answered this question. Of that subset, only 47 percent felt they were completely supported; 32 percent felt they were somewhat supported, but not to the extent they were expecting; and 21 percent felt unsupported. Collectively, a majority of women surveyed (53 percent) felt they did not receive the support they needed to reintegrate into their tech careers.

The process of reintegrating into work after giving birth, and/or after taking a hiatus to raise children, can be stressful. This transitory phase of life applies mental pressure on women who are already often feeling over-whelmed. Thus, lack of workplace support is another common catalyst for feelings of guilt, defeat and even self-doubt.

Have you ever struggled with mental health while working in tech?

Out of the 318 respondents, 184 mothers responded that they have struggled with mental health (a 59 percent majority). Given the stigma attached to mental health issues, discussing the impact that this figure has upon the industry as a whole is a healthy way to change the narrative.

When 59 percent of a professional demographic

struggles with mental health, it is a sign that the industry as a whole has an issue. The irony however is that if mental health is discussed, it can result in unfavorable treatment. However, it is often the unfavorable treatment that causes stress and feelings of worthlessness. While cognitive health takes its shape from natural biological traits and personal experiences outside of the proverbial workplace, the discrimination, harassment, sexism and unsupportive work practices will most certainly influence a working mother's mental state as well. This makes it difficult for new and aspiring mothers to integrate into tech, and can even push them away from the field altogether.

The ramifications of these issues do not remain on an individual level. Rather, as women withdraw from the tech sector, they take away all of the potential contributions they otherwise would have brought to the tech world. For the future of the global tech industry, this can be disastrous, especially when considering how many achievements women and mothers have already brought to the field.

Addressing mothers' concerns within the tech industry and creating an environment that enables mothers, and all underrepresented groups, should be the focus of this new post-pandemic era. Let us provide women with utmost support and change the culture for the better, and that means focusing on the well-being of everyone involved. Thankfully, there are many mentors and sponsors who have supported mothers in tech and paved the way forward for them. The need is continued — let us all step up and assist aspiring mothers in their tech journeys.

Do you have a career sponsor or mentor?

Of our respondents, 39 percent of mothers said they had mentors and sponsors throughout their tech journey. While that number is abysmally low, it still provides hope. Women we surveyed had people who were willing to step up for them; however, when 61 percent of mothers have to take on the industry without any guidance, there is definitely room for improvement. As mothers within this book have emphasized, women should stand up for other women and work towards the common goal of empowering mothers in tech. This will give the best possible opportunity for women thinking of joining tech, and for those that are already here, a chance of achieving success in their careers.

About the Survey Respondents:

While we have discussed the survey results, it is just as important to discuss the diverse individuals who

EXPERIENCES TO INSPIRE YOU   133

contributed their valuable input. Some of the mothers in our survey have chosen to keep their information confidential, and we have respected those wishes in the reporting of our data.

Sexual orientation: Of the 318 responses collected, 93 percent disclosed this information. Among these mothers, 277 were heterosexual, nine bisexual, three lesbian, three pansexual, two queer, and one asexual. The disparity between heterosexual mothers and mothers with other sexual orientations does make us wonder how heavily sexual orientation affects career prospects as a mother; this is note-worthy as there is limited evidence showing that sexual orientation is correlated with outcomes as a tech worker. Gender, color and sexual orientation should be agnostic traits for enabling and evaluating mothers in technology, however more research is needed on this topic to understand and formulate hypotheses.

Marital status: Of the 311 mothers who responded to this question, 88 percent shared they were married at the time the survey was conducted; seven percent were divorced, five percent were single and one mother was widowed. While all of these wonderful women have persevered in the face of chal-lenges, we have to really appreciate the 37 mothers who made their way in tech as single moms, as raising a child alone while working a career in tech is definitely a challenging and daunting task.

Education: All 318 women shared they had completed high school, six women (two percent) had associate's degrees and 109 (50 percent) had a bachelor of arts or above. In terms of higher education, 127 women (40 percent) had completed a master's degree and six women (two percent) had a doctorate.

Another inspirational statistic regarding education is that out of 318 mothers, 69 were the first in their families to obtain a post-high school education, which shows that mothers are stepping up and breaking through the norms to build their own success. We truly hope that more mothers pursue their dreams and passions while inevitably changing the world.

Age: The majority of women fell within the 30-49 age bracket, with 52 percent of mothers between 30-39 and 34 percent of mothers between 40-49. There were eight women (2.5 percent) between the ages of 18-29 and 34 (10.6 percent) of women between 50-59. Only three women in our survey were 60-69, and no one above 70 answered the survey. These results show that a majority of mothers in tech are middle-aged.

Considering age, it is also essential to discuss how much time mothers have spent in the tech industry. Thirty four percent of survey respondents (109 women) shared they had been in tech 11-20 years, 34 percent stated that they had six to 10 years' experience and 14 percent listed less than five years' experience. Sixteen percent of mothers surveyed had between 21-30 years' experience while only eight women (three percent) had more than 30 years' experience.

PART 4

OVERCOMING UNDERREPRESENTATION

ELEVEN

INVEST IN YOUR KIDS

LAUREN'S STORY

> "For the first time in my career, I left my role without a next job secured and thought, 'Okay, I'm going to get to the bottom of this and help create a framework for companies and people to be more successful at integrating technology—allow them to thrive rather than just be productive.'"
>
> —Lauren Kelly, CMO, Thought-Exchange

How did your career first begin?

I started my career more than 25 years ago, but did not actually enter tech until more recently. I pursued psychology as a major during my undergrad and was also intrigued by organizational dynamics. I felt excited and in my element when I landed my first job at Monitor Group (now Deloitte), as a management consultant. This gave me the opportunity to work closely with diverse clients across a wide range of industries—from pharmaceuticals to beverages and spirits to deregulating utilities—helping them solve complex problems.

Over the four years, I began to recognize that there are patterns that actually transcend businesses and organizations, and that there are these transferable, almost universal insights that you can borrow from one industry and take to the next. This made me start to think that as I forged my career, I would not necessarily need to be confined to a certain "domain set." In fact, I started to hone my value as an individual who could bring a fresh, non-industry perspective to the table.

I also learned from my experience as a consultant that while advising companies was really energizing and great for learning, it was not as fulfilling as I thought it would be. I hoped it would be even more rewarding for me to be fully integrated into an organization and more connected to its long-term results, both in terms of financial upside and also feeling that sense of pride in the outcome. So, I decided to round out my on-the-job consulting training with a more formal business education and went back to Harvard to get my MBA. After that, I joined PepsiCo as part of a long-term general management development program in which you are given very large-scale assignments with a lot of responsibility. I spent five years traversing PepsiCo's vast portfolio of mega brands in food, snacks and beverages. I worked across global markets and across marketing, sales and strategy functions. My goal was to learn and do as much as possible with this incredible opportunity I had been given. The last role I held at PepsiCo was leading out the strategy across the company for our Americas food and snacks portfolio, which was a big, profitable and multifaceted business involving a lot of local acquisitions. Through this role, I got a comprehensive view of the global nuances that impact food consumption and production. I was inspired by how the diversity of global palates and culinary customs reflected the amazing diversity of people.

When did you finally switch to tech?

After almost eight years at PepsiCo, I started to wonder whether I needed to more deeply understand the massive impact that technology was having on the internal workings of business and the external expectations of customers. By staying in food and beverage, as fun and timely as it was, I was restricted solely to the leading edge of popular culture. I worried that I could be at a disadvantage from a career perspective if I continued to have such a narrow focus and if I did not find a way to upskill myself and step out of my comfort zone. These thoughts coincided with a call from Dell to take on a similar strategy role to the one I had been performing at PepsiCo. So I traded potato chips for microchips, and that was my first foray into the tech world. It was a very different type of chip with which I lacked experience, so the initial process was uncomfortable for me. In my previous roles, I was the executive who was always on the phone with tech support. So when I took the leap into deep tech, it was not an obvious move to many around me.

How did you adapt to the change in environment?

Before I was hired, I talked to the leadership team at Dell and asked them what their expectations were. They explained that they were looking for an enterprise executive who embraced the customer's perspective and had led organizations through uncharted transformation. Dell was on a journey to expand beyond devices and point solutions to deliver enterprise solutions. I told them that "uncharted change" was my speciality and took on that ambitious and ambiguous role at Dell. So I went from a business that I knew like the back of my hand—the snacks and beverage industry, which was very accessible and did not require a lot of technical depth—to a place where I would likely never grasp all of the ins and outs. One thing that helped me adapt in this scenario was leveraging the knowledge and insights of internal subject matter experts, while in tandem providing principally non-technically focused frameworks for the organization to deliberate on the transformation.

When did you have your first child?

My initial few years at Dell involved extraordinary growth. It was also around the same time I was experiencing tremendous growth in my personal life, because I had my first child and needed to adjust to being responsible for a new life while maintaining the standards I had established at work. And it was at this time that the company went private, which led to tremendous transformation within the organization. Having these changes to my personal life as a new mother and professional life as an executive with a now private company could be received in two different ways. Be overwhelmed or reimagine what's possible? I chose the latter. In short, what got me to this point wasn't going to get me to the next place in my career. When the company began to change, I decided it was a timely opportunity to delve back into some of the marketing strategy work that I loved. I also took the opportunity to step out of the weeds a bit and trust my amazing team to deliver. I sharpened my focus on leadership efforts that would ensure my team's hard work delivered meaningful impact for business.

How did you deal with your responsibilities as a working mother?

It was right around this time that I was offered the opportunity to become CMO for a large real estate development company in California. At Dell, I'd been feeling an itch to get back to marketing again and personally was also open to a change in pace, with more flexibility to balance the requirements of motherhood with a thriving career, so moving to the west coast was attractive and I took the job. Sometimes a role or company change is perfectly timed with a family change. It allows you to reset expectations of yourself and of your company on the way in.

This ended up being an interesting move, because a lot of the goals of my new company were around how we embed technology and combine physical and digital to create experiences and environments that are really fulfilling for people. We had vast holdings in retail, residential, hospitality and office properties. I focused my efforts on helping large enterprises differentiate by creating workplace communities that would foster high productivity and great collaboration. I stayed in this role for almost six years. As CMO, I really shaped and transformed the product and how we marketed it— helping both become more digitally integrated to appeal to the next generation of professionals and leaders. And then COVID happened. Overnight, the business imperative of having a physical workspace evaporated. Across companies, we were all being required to do our work in a more digitally-connected and technologically-enabled way. At the same time, the pace of business change brought on by the pandemic required people and companies to make decisions faster and be more agile. I knew immediately that with this massive shift, I had to get back to technology. And it was a need more than a want. I had become highly passionate about the future of work technology and wanted to make a difference in this realm.

So for the first time in my career, I left my role without a next job secured and thought, 'Okay, I am going to get to the bottom of this and help create a framework for companies and people to be more successful at integrating technology—allow them to thrive rather than just be productive.'

As I was researching companies and becoming more knowledgeable in the area, the company I am at right now, ThoughtExchange, came onto my radar screen. At ThoughtExchange, we believe that organizations can be orders of magnitude more efficient, effective and equitable with the power of collective intelligence technology. The universal shift to hybrid work opened a once-in-a-generation window for companies to usher in a new era of empathetic tech-enabled leadership. I have had the privilege of working with some of the best-known and respected organizations around the world to help them navigate our ongoing shift to a whole new (better) work world whether hybrid, remote or any other form. I made a huge leap going from large, multibillion-dollar enterprises to a high-growth SaaS startup. I am glad I made that decision to return to tech, because it allowed me to provide for my family and the needs of my children while making a difference in the world.

Did you have more children during your career in tech?

I had been in my prior CMO role at the real estate development company for about a year or two, and my husband and I decided it was the right time to grow our family. In trying for one more, we hit the "Daily Double" and were blessed with twin daughters. While I have essentially blocked out the sheer exhaustion of that first year, I am quite thankful for the efficient way that our family expanded. Having children close together, the three of them all being within two-and-a-half years of each other, really helped them bond. I can see they are a tight pod. They are so supportive of each other, and I am excited for them to maintain that relationship in the future. I have three daughters, and I love that they are all girls.

Having twins while I was the CMO of a busy company with a lot of responsibility was a considerable challenge, especially because the real estate company I was working for at the time had never before had a pregnant female executive at my level. They had female executives who had grown children and many male executives, but they had not yet experienced a female executive actually going on maternity leave. I was in the position of having to define for the organization, 'What would success look like?' What would it look like for us to support the business and support me during that time? One of the things that having twins at this company taught me was that I needed to be much more vocal. I was compelled to stand up for the things we needed as mothers, because a lot of the policies and the processes weren't set up. We had to build out a mother's room, retool maternity leave and think about better accommodations for moms returning from leave. Several people on my team subsequently had babies and appreciated that I had paved the way so that they did not have to handle those difficult conversations or advocate for their rights when they didn't feel as secure in their standing. I personally did not face major setbacks, as I was working in a senior position and making key decisions, and my experience with my first child informed me of exactly what I would need to fulfill my role as a mother while ensuring peak productivity within the organization.

Now, as an executive at ThoughtExchange, I am impressed with how my new firm has made inclusion so central to everything we do. As a working mother, I feel incredibly supported. Even more, I feel the latitude to support and celebrate my team members in living their personal lives fully. We work hard, but we keep things in perspective and encourage our people to prioritize family and personal well-being.

What were your biggest challenges and what advice would you give to women in tech thinking about having a child?

So, I understand that my words come from a privileged perspective—that I had all my children when I was in a sufficiently senior role to considerably influence my situation. There are many struggles that I avoided entirely due to my position. However, I have also worked closely with earlier career team members to help them better integrate work and family. The first thing is to identify other people who are more senior and have gone through that experience. After identifying them, it is crucial to ask for advice—and believe me when I say this, because I was someone who probably did not ask for advice enough and wish I had. I might have been able to avoid my own missteps.

The other big obstacle for women is we tend to assume the weight of the world on our shoulders and take everything on. We as women, particularly in business and then technology, want to see ourselves as capable and independent. However, there is no harm in external guidance and help. This was a concept that I had to accept and become comfortable with—delegating a lot at home and at work, really stripping down the things that I was going to personally do to those that added distinct value. Accepting this mindset truly helped me to become a better mother. Initially, I was the person on the weekends cleaning every crevice of my house and doing grocery shopping because I somehow thought that it was going to define me as a more well-rounded person. However, I realized that my time had a lot of value. If I had to choose, I should probably be choosing to spend time with my kids versus doing these other tasks.

My last advice is rather difficult, but ideally you choose a partner who is truly committed to being a co-parent or choose to raise your children alone. As much as we like to say that success as a working mother is about being in a certain industry, company or role, I think much of what makes working motherhood work is by having the right support network at home. That includes a supportive partner, if you choose. It also includes a set of people who you can hire to help you sort out things and save you time (Instacart, anyone?). It includes being near family and good friends. Initially, this might make you feel a bit vulnerable, because you are taking help from others, but it is 100 percent essential. I have a mutual village of people where we collectively help make each other's lives possible.

What We Learned from Lauren:

The most essential lesson from Lauren's story is how the greatest supporters of women in tech are other women. Lauren herself is the biggest example of this, as she helped set up precedents and systems that allowed other women to embrace motherhood during their tech careers without an adverse impact on their lives or the lives of their children. Lauren also showed us just how crucial it is for mothers and prospective mothers to support one another, so that women will not be hesitant to embrace motherhood for fear that it will make their careers suffer. This also encompasses standing up for mothers or potential mothers who are unable to stand up for themselves, at any level within the company.

Lauren also teaches us just how important it is to be vocal. To make a concern heard, it needs to be brought to the attention of the respective individuals. This is only possible if women stand up for themselves and make their rights and needs known. Women everywhere need to raise their voices and make their concerns known so that they can be addressed properly.

Another crucial concept Lauren teaches us is to ask for advice. Refusing to seek help or alternative perspectives might be something mothers often do in order to maintain a reputation of capability, but failing to ask for advice is often a misstep that can lead to unfavorable conditions. While an individual is often considered lacking if they need to ask for advice, it is actually a great thing to do in the long run, as more perspectives allow for more well-informed and appropriate decisions.

Additionally, Lauren highlights her own willingness to outsource things as a key to her success. As a working mother, there will often be times when the workload can interfere with family life. For this purpose, it is crucial to keep a list of contacts who can fulfill work responsibilities for you while you dedicate due time to your children. Outsourcing work and other projects can free up a significant amount of the day and reduce responsibility; in turn, greater time can be dedicated to family and the children at home, which will ensure a proper work-life balance.

Lastly, Lauren reminds us of the necessity of developing a support system so you can have someone to fall back on for support in your career. This system can be limited to a partner or spread out over a vast number of family members, friends and associates.

What Tech Leaders Can Learn from Lauren:

Lauren's story clearly shows us just how capable women can be within the tech sphere, even if they have the additional responsibility of motherhood. Therefore, tech leaders need to understand that women and mothers can be strong additions to a company, and that motherhood does not inherently cause adverse effects. It is crucial for tech leaders to understand this, as the general perception is that mothers can't be effective professionals, which is clearly untrue.

Another key takeaway for tech leaders is that it is essential to create an environment where mothers can approach the management and take up their problems, irrespective of their designation or title. Lauren's observation that women in lower positions often avoid bringing up their problems should be a slap in the face for the industry, because if women feel that they will remain unheard or be ostracized for their opinions, then the industry is not inclusive. There should never be a situation where an individual feels that their words hold no weight.

Tech leaders must also understand that tech is taking over the whole world. As Lauren put it, tech has become a force that is constantly changing and being integrated into the world at large. Therefore, tech leaders must consider their businesses essential and take steps to enhance the quality and efficiency of their business activities. The primary step here is to ensure that the workforce is strong and capable, and many industry leaders usually hold this perspective about men, which leads to male dominance in the industry. Tech leaders must ensure that women, and particularly mothers, are given similar chances and are not discriminated against due to preconceived and often wrong-headed assumptions. Tech leaders must give everyone, including mothers, a chance to prove themselves, and acknowledge that gender itself has nothing to do with an employee's performance. They must stop stereotyping and adopt inclusive strategies—this is crucial for the effective, continued operations of the industry.

TWELVE
BEING BOLD IN TECH AND BEYOND
JODIE'S STORY

66 "Ask for the things you want. The problem is that we, women, do not understand our value."

—Jodie Davies, Founder of Next Session

The woman in this next story endured years of difficult situations, passive-aggressive working environments, and suppression of her true self while working in tech as a mom. Well before her foray into entrepreneurship, Jodie started her professional career working for the global consulting firm PriceWaterhouseCoopers (PwC) where she became the Director of Mergers and Acquisitions in the information technology (IT) space. Despite her fondness for the role and the company, Jodie and her wife had family plans that precluded her from continuing the pace of travel required for her current job. Instead, Jodie took a new role at Mattel, where she was the Director in charge of the company's governance department where she oversaw IT projects and investments. During her tenure, Jodie and her wife had their first child, a son named Lochlin. Jodie and her wife DeAnne grew their family with the addition of a daughter named Kenzie. At the time, Jodie was serving as the Vice President of the Project Management Office (PMO) at Guthy-Renker, and then at its spin-off SaaS company, OceanX. Now in her current role as the founder of Next Session, Jodie has prioritized her family and her other passion, protecting our planet. She resides in El Segundo, California with her wife DeAnne, and son Lochlin (11) and daughter Kenzie (9), and dogs Mac and Lolo.

What were some of your greatest challenges while working in tech?

At the start of my career, everything was professional and I played the fancy corporate part with my laptop and my suit. Later in my career, I became a mum and was uncertain about how to navigate that while working in a male-dominated field. At one point, there was a great switch in my demeanor and I enjoyed it because I felt like it was not just my professional self at work, but I, as a whole, was there. I got to be *me*. Because when you are wearing jeans and Vans, and everyone else is dressed up, people are going to look at you and say, 'Oh, I did not think that you are a Vice President. You do not dress like one.' I would always say, 'I do not need to dress like one to be one, I dress like one because *I am one*.' I will never back down from that mindset because this is who I am.

As a mother, a woman and a lesbian, how did you adjust within the tech environment?

As I matured, I eventually adjusted my attire to suit me rather than the norm. I am a gender-neutral to slightly masculine dresser and that helped me navigate the dynamics of mostly men in the room. I can be the only female at a meeting and not be viewed as stereotypically female. That allowed confidence and trusting myself to create my own path forward. However, when I became a mum, that is when I really felt the difference.

Like many mothers, I found adapting to parenthood difficult, especially since I worked with colleagues and leaders who were childless. They never really understood when I said, 'The kids are sick, and I have got to go and get them.' There were always consequences when I had to do that. I would get these sarcastic remarks and was the subject of passive aggression. If I asked a question, I was told 'If you had been in the meeting you would know the answer.'

I recall feeling overwhelmed and frustrated at times. I took the time needed to care for my children, but I would be lying if I claimed that it did not worry me or stress me out. I eventually realized that the work environment was in discord with my principles and I moved on to find a company with a team and culture that was a better fit for me as a mum who likes to wear trainers and have my hair cut as I liked.

How did you balance the many roles you had as a mother and a tech worker?

There were always consequences. I would get these slightly sarcastic remarks about being in a meeting, and there was passive-aggressiveness in the tone of my superiors whenever I talked about these requirements, so there was always a barrier there. It was hard mainly because the people I worked under had never experienced that life, so they did not really understand the struggles behind it. Things that should have been straightforward for me as a mother were actually very complicated.

What advice would you give mothers and

women who are looking to start a career in tech? What limitations does tech impose on a woman's choice to become a mother?

I think the very first thing is to find what you love doing. You have to really reflect on your own skills and see where your passion lies. If it aligns with tech, then you have all the opportunities to do well within the industry. More importantly, you need to really find the people and companies that suit your authentic self. Even if you are working in a big company, it is the team that you are working with that can define your habits and help determine your happiness level, so it is mostly about the people. At the same time, if you need the flexibility, you need to look at the policies and benefits that companies are offering and make your decisions based on that.

While most companies are still not where they need to be, if you are uncompromisingly honest with yourself about your personal and career goals, and do not put one over the other, you can thrive. It is your life, it is your career, and no one is going to care about it as much as you do, ever. You have to own it. You have to be really balanced regarding the whole thing, so if you find yourself saying, 'This is a great career, but it is going to take too much of my life,' or if you find yourself saying the opposite, you have to take a step back and really decide where your priorities lie. That is how you can make it work.

What We Learned from Jodie:

When a person in the workplace is the only one of their race, gender, sexual orientation or other personal classification, global consulting firm McKinsey & Company refers to them as an Only. "Women who are 'Onlys'—they are often one of the only people of their race or gender in the room at work—have especially difficult day-to-day experiences. 'Onlys' stand out, and because of that, they tend to be more heavily scrutinized. Their successes and failures are often put under a microscope, and they are more likely to encounter comments and behavior that reduce them to negative stereotypes." Furthermore, a person like Jodie who has multiple "only" characteristics (being a female, a mother and a lesbian) is referred to as a Triple Only. "Being an Only or Double Only can dramatically compound other challenges women are facing at work. Mothers of young children are one example of this— they already face more bias and barriers than fathers and women overall, and when they are often the only woman in the room in their workplace, their experience is even more difficult." Recognizing how women like Jodie overcome many challenges that exist just for being true to themselves is one such way that we hope this book inspires others.

A key takeaway from Jodie's story is just how important it is to maintain a strong sense of self when working in the tech industry. Jodie has proven that a person can work according to their comfort level while still being as capable as anyone else. It is inspiring to know how Jodie challenged accepted norms and bent them according to her vision so she could do justice to both her professional life as well as her personal life.

She also shows us that "real life" as mothers can present logistical challenges; however, the solutions to those challenges are unlikely to affect performance or productivity. Jodie has, just like many of the other mothers within this book, emphasized the need for women to stand up for themselves and be vocal about what they need and want. She teaches that it is perfectly acceptable to fight for what you need, because at the end of the day, an individual's performance is heavily dependent on their comfort. She also shows that it is necessary to have a clear goal and mindset regarding a career in tech, and that mothers or anyone working in the field need to have their priorities in order so they can properly divide their time and attention towards their professional and personal lives.

Finally, Jodie's observations also emphasize the need for more trust in the workplace for working parents—a mutual understanding that working women are adults who will be held accountable if their deliverables are unmet. Until that day comes, managers and leaders should let moms go to the school plays, volunteer on the school board and take their children for after-school ice cream. By letting mothers show their children that they have balance and are committed to both loving them and thriving in their jobs, together, tech companies and mothers can pave the way for a true future of change.

What Tech Leaders Can Learn from Jodie:

Tech leaders need to realize that the environment they create for individuals can greatly affect their performance and their work. Leaders must realize that if they are promoting a culture of strictness and intense professionalism within a company, it will always come at the expense of comfort, which can lead to low employee productivity and satisfaction in the long run. Therefore, it is essential to create a flexible environment where individuals can focus on their work in a comfortable environment while also having the ability to tend to urgent personal matters. This will ensure success within the company and guarantee that employees can perform to the best of their ability without having the stresses of their personal lives on their minds. Leaders must also be mindful of their employees and their requirements to ensure that women and mothers are not sidelined within a primarily male-dominated space.

Perhaps the greatest takeaway from Jodie's story for tech leaders is that a person's worth is not determined by their appearance or the way they present themselves. Every individual brings a unique set of skills, and those skills show in their work rather than their appearance. This is evident from Jodie's appearance as a vice president, where she went against accepted norms and still performed excellently. Tech leaders need to give their employees this flexibility to ensure that they are comfortable enough to put forward their best effort. Proper implementation will guarantee that they continue to perform to the best of their abilities and contribute immensely to the success of the company.

THIRTEEN

BREAKING BARRIERS

> "I consider barriers as ceilings, and there are a number of them for me that I have to push through. Using those barriers to drive change is necessary, and at the position I am in, I feel that I can play a much greater role in that. I do not want others to go around those barriers; I want them to push them over."
>
> —Tracy Taylor, Markets and Alliance Lead Partner at PwC, New Zealand

Just like life experiences are unique to each person, so is each person's career journey. While Jodie Davies's gender-neutral or masculine attire and persona gave way to inclusion in the boys' club, Tracy Taylor, unfortunately, had a less favorable experience.

Tracy has spent more than 20 years in the tech and management consulting business and has learned a lot from her experiences. Tracy is a mom to three children (six-year-old twins and a four-year-old) and a puppy, whom she raises with her wife, Claire. In her current role, Tracy is running the Markets and Alliance business, which effectively means that she manages risk, management and technology consultancy for PwC in New Zealand.

Like others in this book, Tracy acknowledges that there are challenges in the way, and there are always barriers to progress. While she is still navigating difficult situations, she is also a shining example of how to be brave, strong and determined, and how to relentlessly work to lift underrepresented groups up from the shadows.

* * *

What are key challenges that you have faced in your career thus far?

There are always barriers, and I have never really overcome them. In fact, I consider barriers as ceilings, and there are a number of them for me that I have to push through. I am always one of the few voices in the room when it is time to talk about issues that women and mothers face in the workplace.

How has being gay affected the way you are accepted in the workplace?

It is one thing I have always been proud of, to raise a family as a gay mother, but there are always setbacks associated with it because the other mothers treat me differently. A lot of that has to do with the fact that they are not able to perceive that I experience the same challenges that they do. When they go home, and I go home, we are all having to deal with the same struggles regardless of my orientation. Very few people can accurately grasp this.

I have learned to [ignore other people's lenses], because there is something really invigorating when I am being authentic about who I am. It is more about transformation than technology, because tech is just an enabler of transformation. That transformation is critical, and it comes through being authentic and open about your journey. It also helps others to perceive the industry the right way, because when you're a senior leader with 20-plus years of experience, people either think you have come from a place of privilege or you have achieved success. It is really important for them to acknowledge the challenges and know the struggles they will have to face to achieve that success. My voice can help show these women and mothers what they can do, and I believe that is very important. Using those barriers to drive change is necessary, and in the position that I am in, I feel that I can play a much greater role in that. I do not want others to go around those barriers, I want them to push them over.

That is really the kind of alliance that needs to exist between women, especially senior women in tech. The conversation should not be limited to the work-space, either, and we should always think about how to support one other even when we are outside the boardroom.

What are some key strategies you use to navigate this difficult climate?

It is important to have female mentors who are visible and stand out. It also helps a lot that our candidates see a woman leading the team, and that goes back to connecting with the same idea that sometimes you are the only voice in the room and you have to speak up.

How do you balance your life as a mother with your career?

When I am at home, I do everything that needs to be done as a mother. I am always open about it, and I do not feel like I have to apologize for wanting to take time at least once or twice a week to pick up or drop my kids off at school. I do not have to ask for permission, and this is something I talk about in front of my team so they know it is something that I am trying to normalize. We need to make it acceptable to balance your life with your work, whether it is regarding children, family members or your own personal dynamic. Gone are the days when you have to work 15 hours to prove yourself, because that is something out of the '80s mindset. I always feel that if you are really putting in effort, you can give your best for six to seven hours a day and that is better than working like a robot.

I am thankful that more and more tech leaders are working towards this goal, and one of them I would like to name is Arianna Huffington. Her books taught me so much about how to truly be a leader, and one of my favorite takeaways is her concept of meetings where we do not sit down to read status reports, and instead work towards more meaningful interactions that actually make a difference. I would say that things are a lot better now, but when I actually had my children, it was a lot more challenging.

As a biological mom, and I believe this applies to adoptive mothers and surrogate mothers as well, I had to have conversations at work that I did not really want to have, because it was something very personal for me. Balancing my work life with my children was a considerable setback. At that time, I was with an organization that wanted more and more of my time, and I worked under a woman who did not really have any of the qualities a leader should. She made it impossible for me to take time off with my kids, and I did not get parental leave until they were 14 months old. It was great to have three months of parental leave at the time, but I wish I had pushed harder to have my right to take it when they were younger. I took the same leave when Harper was born, but I was the first woman under the same-sex parenting rule to take parental leave, and it took a while for the organization to wrap their heads around it. But I personally feel that taking that step was a catalyst for change, and it was really important for other people to realize where they stood and what options they had available to them, and I personally feel that it had a positive impact on the company culture as a whole.

Do you feel that women and mothers have to go the extra mile to fit into the tech space?

There are several milestones surrounding the

image of a woman in the tech industry that I would like to discuss. The first is my own experience, where I have always had the voice and the confidence, but have felt lacking when it comes to appearance. I've had to battle imposter syndrome, since I do not typically fit the ideal image that an organization expects of their female employees. I have always had the mindset of putting my comfort over what is expected from me, so I will wear dresses when I want to wear dresses and be as comfortable as possible so I am in the right state of mind to work. There has always been a struggle of gaining weight, losing weight, dressing a certain way just to fit in, because I want to look like the other women, and that is a thing that has been a constant from the start. This fragility is not something that is limited to me, either, because I have seen it at every level within the female workforce.

I have seen employees who have not come from a wealthy background wearing borrowed clothes, and it is so heartbreaking to see someone wearing an item of clothing that does not fit them, and you can see the discomfort they are facing as they try to blend in. I think we need to eliminate that mindset, that you have to look or dress a certain way to be a part of the organization. I have had my struggles with the idea, too, and there was a recent experience where we shot a video and I did not really look my best in it. I thought about canceling it and doing retakes, but I eventually gave my approval for the publishing, and it was actually very well-received. I can never watch it, though, because when I look at myself, I just do not feel like I fit in. No matter what I have done or how much weight I have lost or how I fit, there is always that imposter syndrome that puts me off from all of it. I am constantly reminding myself that it should not matter, and when it comes time to dress up, if a pair of jeans and a dress is comfortable enough, I will force myself to pick that outfit.

It is not always a straight slope, because I avoid wearing heels and I have been made fun of by male co-workers joking about my 'flat lesbian shoes,' so those taunts are always disheartening. But at the end of the day, I have learned that I would rather be comfortable, and I am constantly convincing myself to ignore all those rude comments and dress up for myself.

Have you ever had a mentor? What advice would you give to mothers seeking mentorship or support?

I will start off with a story from when I was 16, and at the time, I worked a government job. I did not have a degree yet, so it was an entry-level job, but I did not get promoted when the others did. And my first thought was to leave like any typical hot-headed teen would, but then I had a word with my boss, who was a female, and I asked her what went wrong. She sat me down, and once my tears had dried up, she told me that nothing is a right and you have to earn your promotions. She said I was great at what I did, but I never talked about my victories. She brought up examples of other people and explained how even if they were worse at their jobs, they spoke up about their work and they advocated *for themselves*. It took me a while to comprehend what she had said, but I understood what she was explaining eventually.

Basically, she was telling me to build those upper relationships, and I tried to follow with what she said by building a strong connection with everyone, even the lady that brought in the tea trolley. Her words really stuck with me, and even after almost three decades, I do not think I advocate for myself as strongly as I should, but that conversation just returns to the imposter syndrome concept where you find yourself questioning if you should be more vocal about yourself. I believe it is absolutely necessary to do so, and you just have to push those negative thoughts out of your mind and go for it.

Another thing I have learned is that you do not always have to find a mentor who is senior to you. Your mentors can be your juniors as well; they just need to have the same shared values. And you have to actively seek them out and make connections and request their advice and assistance. In my experience, nobody really turns you down when you are just asking for guidance, so that authentic advice can really make the difference. Once you initiate it, the process is simple and really takes you a long way.

Out of 300 survey respondents, 54 percent have been harassed at work. Do you have a similar experience?

There are some things I would not like to share because they are very personal, but I do have some experiences that I would like to talk about. There are times where I have been bullied within the tech space, and there is a lot of indirect bullying that I would like to talk about. It really makes the imposter syndrome worse, so I feel that it needs to be addressed. One of the things I have noticed is that if you ask for support, you are viewed as incapable and your authority is often taken away on the premise of giving you time and space, but that in itself is a form of gaslighting and bullying where you are sort of pushed away from the work itself instead of being given the appropriate support to see your work through. It is something I have personally experienced and I have seen it happen to other women and mothers, too.

There have also been a lot of comments regarding my appearance and the way I dress, and those are always repulsive. I can appreciate compliments, but when I am asked to dress a certain way to appeal to someone's opinion, it really puts me off. This kind of harassment exists all the time, and the worst part is that it makes you question if you should do things differently when in reality, those words should never be said in the first place. Those comments always make me step back, and that is a problem for the industry as a whole, because we should have an environment where people want to step up rather than step back. There are comments on clothing, hair, makeup, shoes and even weight, and they have always been repulsive in my view. It is the stress of having to conform to ideals that really puts you down, and you are perceived differently based on your appearance. That in itself is a form of harassment, and I have had to deal with that throughout my career. I feel that we as mothers and women need to step up and make it clear that we are not going to accept this form of bullying and harassment. It pushes so many women away from tech, and I do not think that is healthy at all.

What advice would you give to aspiring tech

mothers and women?

I always say that if you are passionate about tech, you should always follow your dream and never take the challenges as barriers, but rather as stepping-stones to overcome. We need more women in tech because we are so underrepresented, and tech itself needs the female voice because women are equally capable and even surpass men in several areas. Their presence is crucial to the industry as a whole, and we need to realize the impact we can have.

At the same time, as women, we need to think about how to enable others around us. As leaders, we need to mentor the mothers and women working under us so they have someone who is prepared to represent them when they are not there and speak up for their rights. It is so important to have that supportive bond between women, because even the smallest amount of time can really make a difference in people's lives. You have to be the advocate for change, and once every woman internalizes this concept, they can truly establish their own worth in the tech world.

What We Learned from Tracy:

Tracy's story is all about standing strong in the face of harassment, bullying and discrimination. She has proven that in spite of other peoples' perceptions, sexual orientation, the way you dress and the norms of the industry are irrelevant when it comes to individual and organizational success. Tracy has proven that you can be an effective and successful tech worker without succumbing to the pressures of misplaced ideas or discriminatory practices.

As she bluntly explains, being a woman in tech has always been a struggle for her, and that despite progress in some areas, she still experiences discrimination attached with being a woman and a mother in tech. Within the population of women in tech, the subset of mothers in tech is smaller—and even smaller (and thus less visible) still is the subset of gay mothers in tech.

Unfortunately, Tracy's challenge fitting into the workspace as an openly gay woman leader is an experience that others have had as well. A 2017 report by the Kapor Center for Social Impact, which surveyed more than 2,000 people who had left a job in the tech sector in the past three years, found 24 percent of LGBTQ people had experienced public humiliation or embarrassment in the workplace. Moreover, 64 percent of LGBTQ employees who were bullied said the experience contributed to their decision to leave.

As women, we should realize that we cannot and should not accept any kind of harassment or bullying, especially when it comes to appearance, as no other individual has the right to pass comments or make statements that make us feel uncomfortable. We need to stand up for ourselves and make it known that we will not tolerate this.

Tracy also shows how important it is to stand up for yourself and for groups that have limited representation. As an individual with an underrepresented sexual orientation within the tech industry, she has often been treated unfairly or ostracized by her peers, though she still stood strong and never walked away from tech. We can all definitely learn from this. Women know that there will be challenges that might push them away from tech, but as Tracy says, we need to step over those barriers and persevere so we may become leaders and allow the next generation of tech moms to achieve success in a more welcoming and holistic environment.

As several other women noted as well, another important thing that Tracy's story teaches us is to advocate for yourself and ask for advice. She's taught us that advice can get us places we were unable to reach before, and finding mentors who can guide you through the process can make things so much easier. "Nobody really turns you down when you're just asking for guidance," Tracy says, "so that authentic advice can really make the difference."

Finally, the most unique advice from Tracy is to focus on meaningful interactions. These can really help build bonds between employees in an organization, which can, in turn, build a support group that is always ready to help its other members in need. Focusing on relationships and interactions also raises awareness for the struggles that mothers face, and can greatly limit harassment by giving others essential insights about the effects of their words. We personally believe that this is crucial in every workplace, as building stronger relationships ultimately leads to better working environments in any context.

What Tech Leaders Can Learn from Tracy:

Tracy has given a clear message when it comes to discrimination within the workplace: tech leaders need to understand that discrimination on the basis of gender, orientation or any other aspect for that matter is unacceptable. A person's worth in an organization should be judged on the work they do rather than unrelated personal aspects such as their sexual orientation. Tech leaders need to ensure that nobody in a company is treated as an outsider; instead, norms should be amended to allow many kinds of individuals to seamlessly integrate into systems and cultures with the same opportunities as everyone else. Tech leaders also need to learn to take a stand against bullying and harassment and ensure their employees are comfortable in the workplace. Only with the proper level of comfort can employees perform their best, so it is in tech leaders' best interest to ensure that their employees are not targeted on the basis of gender, sexual orientation or appearance.

Additionally, tech leaders should also realize that they should offer support when employees reach out for guidance rather than distancing those individuals from their work. Instead of considering them a liability, tech leaders should focus on giving individuals who are unable to perform their best due to excess stress the right tools they need to excel and find peace. Taking away authority or limiting someone's decision-making power simply because they need support is in itself one of the worst forms of bullying, so it is essential that tech leaders not resort to such practices. Instead, they should focus on resolving these issues in a way that addresses the underlying cause.

Tech leaders should do their utmost to make life easier for tech moms rather than creating more hurdles. To perform their best, all organizations need to create frictionless work environments and systems. Still, tech leaders have to understand that the biggest role they play in that journey is defining the journey itself.

FOURTEEN

MOTHERHOOD MADE MY SUCCESS

CHRISTINA'S STORY

> "I became more successful in my career after having kids because I learned to juggle motherhood with work, hyper-prioritize my time and figure out where I need to spend my focus."
>
> —Christina Wong Singh, VP Customer Success at Split.io

There is no dearth of Asian Americans excelling in all kinds of jobs in the US, but achieving peak success in tech as an Asian American woman and mother remains a rarity. Still, Asian Americans are not under-represented in the tech world. A couple of years ago, mainly to disprove the assertions that there was any demographic disparity, Facebook, Google, Apple and Microsoft released employee data that showed that

Asians made up at least one-third of each company's workforce. At Meta, up to 52 percent of the workforce is composed of employees of Asian descent.

Still, despite the demographic's strong presence in the tech industry, Asian and Asian-American employees find they are often sidestepped when it comes to reaching the executive levels—especially women.[1]

This is one of the reasons why this chapter is dedicated to the success story of an Asian American woman, Christina Wong Singh—"raised with an immigrant mentality," as she recalls her upbringing. Her story, perspective and life experiences are unique to being an Asian American woman, and yet they are surely relatable to any woman in the tech world today.

* * *

How did your journey into a tech career begin?

I grew up in Brooklyn, New York and I went to NYU and majored in finance and accounting as I was good at math. Stereotypical Asian trait—being good in math. Both my parents were in accounting, so I thought it would be a natural progression, and then I got an internship in finance. All day, I was sitting and crunching numbers. Even though I was good at it and it paid well, it was not for me.

It helped immensely that I was at NYU because it afforded me lots of different opportunities. I knew I wanted to be in business, and I wanted it to be customer-focused. After many different internships, I got a job as an account manager in an advertising agency. That is how I spent the next 10 years—in the advertisement agency world.

Five years into it, I realized I had never stepped out of New York. I lived there my entire life and did my college there, too. I wanted to explore more so I moved to San Francisco in 2005—I had gotten a job offer there.

The ad agency world there was crumbling. There would be headlines in newspapers like 'San Francisco is a dying market,' but I joined AKQA, which is a fully digital agency and I enjoyed myself. Tech was all around me, and I was curious, thinking, 'What is this world of tech? What is happening around me?' I wanted to participate in it.

I had become a people manager in the advertising world but I did not have the experience in tech, so I started as an individual contributor customer success manager, worked my way up, was asked to run customer success, and then took on professional services. Two companies later, I am now an executive driving customer success. My last 20 years have been about creating great experiences for customers, helping them maximize the value of the product or the service.

How did kids fit into your scheme?

It has been a lot of hard work, determination and just following my gut. I always knew I wanted kids. That had always been part of my journey, and in life, you cannot plan those things—you just have to let life unfold. But like I always say, I became more successful in my career after having kids.

I had my daughter when I was 34 and my son at 37. I learned to juggle motherhood and my career. I learned to prioritize my time and my focus—and that helped my career take off.

In what other ways did motherhood add to your professional success?

Motherhood taught me empathy and authenticity. It brought about an understanding that we might come to the office and do great work, but we have other things happening behind the scenes in our lives, too! I mean, things could be crumbling in the background, but we come to work with an attitude that everything is all right.

I also learned that people are scared to share their stories about what is going on at their homes, so it is about getting to know people and checking in, letting them know it is okay to take the day off sometimes. It has taught me that sometimes I have to bring the baggage to the office, but still do amazing work. That way, especially in tech, we are lucky—we can work remotely and do not have to always be tied to our desks. We can go pick up our kids if we need to and come back to the meeting. It helps us be who we are, so I have a well-rounded balance of my work persona and my personal persona.

What challenges do you face as a mother in tech?

The most basic thing about being a mother in technology, especially at an executive level, is that not a lot of people are mothers. There are not even a lot of women to begin with. And so, the men I work with do not have to deal with their kids all the time, but I am constantly juggling many things. I prioritize my time according to my kids' schedules and the other things happening around me.

I went back to work after four months of having my kids, and I was very happy going back to work. That was not my parents' perspective. They said, 'Why are you going back to work so soon?' There were other issues, but Asian identity urged me to pretend everything was perfect. I woud go to work as if everything was okay. I soon started to realize that sometimes, everything might not be okay and I need to be able to showcase that as well. This has been the most interesting evolution for me.

How has your experience in tech been as an Asian-American woman and mother?

I first think about being an Asian-American woman in tech and then being a mother in tech. I was taught not to question things unless I was sure that something was wrong. So that is something I have had to peel away and learn that I have a voice and I need to speak up, to say what I have to say—that I need to interject sometimes and not be interrupted. That my thoughts are as valid as anyone else's—for me, that's been a bit of a struggle and learning experience because of how I was raised.

How did you overcome this and find success?

Having people in my corner reminds me not to have that complex. I have to remind myself, 'This is where I want to go, and for me to get there, I have to step out of my comfort zone and speak up.' I have a daughter and a son, and as much as I try to make sure the gender lines do not go, 'You're a boy, you're a girl, this is how you act and behave,' I realize that I do treat my daughter differently than I do my son, sometimes. That is also who I am—it is what is ingrained in me. Above everything else, I want to teach my daughter that she does not have to be perfect. That word is something that I was taught, and for me, one of the biggest moments was realizing that I have to be an example for her. She loves to play sports with boys and other girls, and I tell her if someone says something to you or hits you, you need to say something back. It does not mean you have got to hit them, but you must speak up for yourself. In a way, this neutralizes my growing-up years and validates the fact that I have been more successful in my career after having kids.

What do you want to tell women in tech planning on having kids?

I say that if your desire has always been to have kids, do it. You cannot plan these things, life just needs to happen. I am a planner, used to like to plan out the weekends—I tell the kids, 'This is what we will do at 10 am,' or, 'This is what will happen at 11,' and that never happens. So I have had to learn to go with the flow and just be in the moment. If those are the things you want, just go for it and it will all work out in the end.

I know there are moms who say, 'I do everything for the kids and I take care of myself last.' But my take is that if I do not get enough rest, I will not be the best mom to my kids. And so, when I put my kids to bed, sometimes I go to bed at 8 pm, sometimes at 11. If I need that time for myself to just rest and recharge, I put myself first so that in the morning, I am not bitter about the time I spend with my children.

Do you have any words of advice for mothers in tech?

It has to be about demystifying this idea that being in tech and being a mom is hard because you have to be young and agile. Add to it the bro-start-up culture of working in tech. But there are those types of companies, and then, many that are not. So, spend some time understanding the culture of the company you want to be in and see if it's right for you. Often people think tech is about the Facebooks, Googles and Apples of the world, but there are small to mid-size tech companies also. So, educate yourself and find your company.

What We Learned from Christina:

Christina teaches us an important lesson about always speaking up for yourself. She reinforces that you do not need to always be completely sure of whether something is wrong, and it is important to voice your concerns despite the uncertainty. This helps lead to healthy discourse regardless of any issue and can potentially lead to a resolution. As so many moms have pointed out, it all involves speaking up and making yourself heard. Voicing your opinions about rights and requirements can help establish you as a representative of yourself and other mothers, and can help identify problems mothers face that might otherwise be ignored.

In America, ethnic minorities, and women among them in particular, have often been excluded by the bro culture of the tech world. Racist and sexist comments are the norm, and the men making them seem to expect women not to be offended. As one example of this, Chia Hong sued Facebook because she alleged she was asked, "Why don't you just stay home and take care of your children?"[2] She also mentioned in her lawsuit that she was made to serve drinks to her male colleagues and organize parties, and was ignored during group meetings because of her accent.

The truth is that while there have been many

discussions on women and diversity in tech, very little has been spoken about the realities faced by Asian American women in this sector. Clearly, the tech world has not seen enough empathetic leaders who care about diversity and inclusion over and above growth metrics.

Moreover, Christina's story also shows the importance of her own mom philosophy of putting herself first. It might be challenging for mothers to act "selfishly" when they have so much love for their children, but it is imperative for them to realize that they need to dedicate time to themselves and their own well-being if they want to give due attention to their children. Stability and a sense of comfort are both necessary to be a productive mother and tech worker, so focusing on yourself is the first step to focusing on everything else in your life.

Above all, her experience highlights the importance of being authentic. Instead of hiding your struggles, it is crucial to share them and be vocal about the challenges you are facing. While there might be negative aspects to sharing your struggles, there will definitely be considerate people who will step up to help you overcome them as well.

What Tech Leaders Can Learn from Christina:

Through her own wisdom as a leader, Christina teaches us just how important empathy and understanding are for a leader. A tech leader needs to go out of their way and make the effort to understand their employees on a deeper level, which includes keeping track of their external responsibilities, mental health and challenges outside of the workplace. While this might be challenging, it is necessary to whatever extent possible for tech leaders to be considerate, make exceptions and create solutions for struggling individuals in the workspace. Doing so leads to a greater level of comfort and well-being for employees, which ultimately leads to a healthier working environment and greater productivity.

Furthermore, tech leaders need to give mothers and women the space to express their discomforts and difficulties. Rather than shunning them for being vocal about their challenges, tech leaders should step up and assist in any way possible to make it easier for these women. By doing so, they can help mothers and women feel more included in the workplace while ensuring that they are in the necessary state of mind to give due attention to their work.

FIFTEEN
CREATING SPACES
KIM'S STORY

66 "As a working mother, I could be a better mom too—and it was tough to grasp that concept initially, but I wanted my children to believe that their mother could do anything."

–Kim Damron, President and CEO of Paciolan

Our next inspiring story comes from a woman who has stepped up in the face of sexism and racial discrimination to establish herself as a leader in tech. Her resilience and determination are impressive and serve as a source of inspiration for women and mothers everywhere. Her name is Kim Damron, and she is the president and CEO of Paciolan, the leading ticketing software company in college athletics whose clients include 80 percent of the Power Five schools. It is the second largest ticketing company in North America, just behind Ticketmaster, and Kim has worked there for 15 years.

As a mother, Kim has four children ranging in age from 11 to 22. One of them graduated college during the pandemic and another is a sophomore in high school, while the other two are in seventh and fifth grade, respectively. Kim started her career at Paramount Studios, where she learned a lot from her female boss, who was an expert when it came to management.

"That was the first four years of my career, and I believe it was very sheltered, because when I stepped out of the studio into the environment of corporate America," she says, "that's when I truly understood the imbalance in terms of diversity, especially at the management level."

* * *

How did you find success in an industry where the odds were stacked against you?

The game-changing moment for me was reading Sheryl Sandberg's book, *Lean In.* I do not necessarily agree with everything in the book, but it was a turning point because it gave me a much-needed boost in confidence. I was COO at the time and my boss was about to retire, so the book was the catalyst I needed to step up and fill in that position. I always thought I was not up for it, but when I saw one of my male co-workers looking to fill his place, I thought to myself that if he could do it, so could I. I was struggling at the time since I was a working mom with young kids, and I have to say that working-mom guilt can be a very real thing. But when I got the confidence I needed, I realized that I could do justice to both. As a working mother, I could be a better mom too—and it was tough to grasp that concept initially, but I wanted my children to believe that their mother could do anything. I am proud of being a female CEO with a minority background. That is something I have looked at as my goal and one of the things I have aspired to do. Almost every company I have worked with had a cis-male CEO, so just breaking through that barrier, getting others to see me as a leader and an equal, has always been a challenge.

How did you balance your work and personal life? Did you take a break or withdraw from tech?

I never took time out of the workforce. I had older stepchildren through my husband who I married 15 years ago, and they have lived with us at least half the time we have been together, but I never needed to take time off for them.

I did take almost a year of maternity leave between my three children, but never extended time away from the industry. I have definitely considered it, though. When my children were younger, I was thinking more about survival than my career path, so I was not thinking about taking on more work or pressure. It was the right decision too, because I was not really capable of taking on more. I was hardly capable of putting mascara on, and I drove to work with two different shoes and even slippers. Those first couple of years were brutal, so I was just looking for a job where I could manage both aspects at once.

When I had my first child, I thought there was no way I could manage. I spent the first year saying that I was going to quit and there was no way I would go back, but I am so glad I put those thoughts away. The pandemic itself has taught us that people can work from home comfortably and there can be a balance between both work and personal life. It's really saddening to know that 2.2 million women dropped out of the workforce in 2020, not just because they were in hospitality but because they simply could not manage it all.

There is a need to create an environment where working moms can be successful but still be a part of their children's lives, and this applies to the men as well. There is a time where the men in the workforce say that their kid has a baseball game, and I tell them to go be with their children because I know they will not regret missing a meeting as much as they will regret missing their kid's first home run. It is so important to create a space where mothers feel they can balance work and family.

As a CEO, how important do you consider mental health? Have you struggled with mental health in your own career?

I am getting older, and at 48, you start to notice the effects these things have, so I try to give as much time as I can to look after myself. The last year has been really challenging too, with the effects of the pandemic. I have been on video calls and have seen people struggling. Our own company had to be shut down because we were a non-essential service, so there were so many layoffs that hit me hard, both personally and professionally.

I started bringing in a psychologist to company meetings, and the response was phenomenal. There is a stigma attached to discussing mental health, so it was heartwarming to see people speak up about their struggles and challenges, and it really helped the company environment since we knew what we needed to do right. It was really helpful, even for myself, and the best thing I learned was to go out and get some fresh air every day. The psychologist would give us challenges and push us to really look after ourselves, and I think that was the best thing that happened to the company. It greatly helped with all the stress I was facing, and it helped the company overall because everyone was looking after themselves, which helped them give their best.

I feel that I am the first to say that we are all struggling mentally, and we need help, and it is so important to understand that and just put it out there so there is a plan to assist where necessary. We need to really give our employees the resources they need because I've had so many come to me and say that they were struggling, but the therapy sessions helped them overcome a significant portion of their stress.

I was there to manage all of that. I think another issue that arose from that, and one that applies to a lot of women in tech, is that they simply do not give time to themselves. For the first six months of the pandemic, I was on Zoom calls, and during breaks, I was taking care of the four kids at home. There is a lot of food and a lot of cleaning up you have to do with a family of six, so it was very tiring mentally since I did not have an escape.

Bringing in a therapist helped me as well. I was shown how to catch up on sleep, because sleep deprivation, from childbirth until the present, still remains one of the worst things for women.

What would you advise mothers and women to do to achieve success in their careers?

The biggest challenge we as mothers and women face is the belief that we cannot do it. We believe that we lack the ability to be a good mother and a successful executive at the same time, but that is simply not true. It is not an easy process, and it is not perfect, but it is possible and we need to realize that.

You are going to have bad days, and your kids might have bad days, but your role as a working mother does not influence this. It is perfectly acceptable if things are not always smooth sailing, and there are always going to be ups and downs, but that should never stop you from working on your career.

There is a woman I work with who, until two years ago, kept saying she could not take on more responsibility. I kept advising her to strive for more, and I knew she was capable of taking on more, and now, she has progressed so far in her career that it is awe-inspiring to hear of her journey. I feel that it is always a lack of confidence that holds us back, and it is important to understand that we are capable of juggling these things. I am personally so happy with my decision to stay in tech because it has been better for my children, better for me and better for my family.

What We Learned from Kim:

Kim's story shows how important it is that the workplace is made appropriate for women and conducive to raising a family. In Sheryl Sandberg's book, she writes that 55 percent of college graduates are women, but only 25 percent make it to senior roles, since most women drop out of the workplace if they cannot find a balance at home. As Kim notes in her interview, she can definitely relate to those pressures.

Her story also shows that sometimes, things might not fall into place, and that's okay. It is important to realize that there will always be ups and downs in each journey, but how we tackle them is what's important. As mothers, we need to stand strong when challenges come our way and tackle the lows while cherishing the highs. There will always be another challenge or struggle, but as Kim has shown, standing strong and working intelligently can help you through even the most diffi-cult situations.

Kim also emphasizes the same point that so many other women have emphasized: the importance of standing up for yourself. In Kim's case, she has seen the results firsthand, so we can definitely learn from her experiences. We need to truly build that confidence in ourselves as mothers so we can defend our work and our skills and prove that we are as capable as anyone else.

The importance of mental health is another essen-tial takeaway from Kim's story. As she illustrates, it is incredibly important for mothers to focus on them-selves and not sacrifice their own mental well-being for their careers. If taking time off or simply doing things for their own comfort and peace can help mothers stay mentally stable under the grueling pres-sure of the industry, then it is imperative that they exercise their right to do so. As mothers, we need to do what makes us happy, as comfort and peace can truly help us excel.

Finally, Kim speaks about the importance of encouraging other women to be confident, which is crucial to the industry as a whole. There are so many mothers with amazing skills who are sidelined solely because they do not have the confidence to talk about their work. Encouraging them to speak up and take the lead can be pivotal in their lives, truly pushing them towards the success they deserve.

What Tech Leaders Can Learn from Kim:

Kim's story is one of the most important for tech leaders to hear, as she discusses the stigma around the topic of mental health. It is imperative that industry leaders start treating their employees, and particularly mothers, as human beings rather than machines. They need to consider that each individual has both work-place-related and personal struggles which can impact their ability to be effective at their job. While tech leaders can rarely help with personal struggles, they can definitely lend a hand when it comes to challenges in the workplace. Kim's decision to hire a therapist is proof of this, and the phenomenal response to her decision only solidifies how important it is to address mental health challenges in the workplace. Tech leaders should go out of their way to incorporate mental health discussions in the workplace so that every employee feels their concerns are being heard and addressed. This can greatly improve productivity and ensure a healthy and supportive work environment.

Finally, Kim also notes the importance of creating an environment where parents can maintain a balance

between their work and family. This level of consideration can truly raise employee satisfaction, and the
benefits it can bring for individuals, tech leaders and
the organization as a whole are substantial compared to
any potential downsides.

SIXTEEN

TIME FOR A DIFFERENT CONVERSATION

ROLANDA'S STORY

> "When my children say that they really need me there, I will move heaven and Earth to be by their side. But when it comes to the little things, I do not partake in them and leave room for my girls to be creative there. I know they need their space and downtime without me too, and I personally feel that makes them better people and stronger women."
>
> –Rolanda Jones, Senior VP of Product and Engineering

The next story is from none other than Rolanda Jones, Senior Vice President of Insights and Innovation at a $360 million tech powerhouse in the wellness sector where she oversees the product management and engineering divisions. As a woman of innovation, Rolanda has worked at Yahoo!, Facebook and eBay before finding her "happy place" in her current role. In addition to her technical chops, Rolanda is also a public servant: she is the founding member of multiple philanthropic organizations that support girls and women in STEAM and is an advocate for women pursuing professional ambition. Additionally, Rolanda also served as the mayor of a Silicon Valley city—all while working a day job and raising children!

What are your responsibilities in your current tech role?

I call it product and engineering because it is easier to understand, but basically, I lead two key technical universes at our company. One is our product services division, which includes product marketing, design, market research and user research functions. The team includes technical programs managers, AI engineers and data scientists because we have similar skill sets that we want to keep together as they grow and develop their careers. I am the leader for all the service teams.

The product and partnership mission involves responsibility for all the APIs, which we use to scale our interactions and our connections with several medium and large-scale customers as well as partners across the industry. There is a key engineering function associated with that as well as a key product and selling function, so that is part of my responsibility, too.

How did you get involved in tech? Did you have a prior background or was it purely based on passion?

I actually studied electrical engineering as an undergrad, so I started off in a completely different sector of tech. My first job involved working in satellite communications for Motorola. We worked on a product called Iridium, which was designed to create communication access in remote locations, particularly regions without proper cellular infrastructure. We launched 66 satellites into orbit and connected them to cellular networks to ensure connectivity for people living in remote areas, but the product was a massive failure. The major reason was the cost—25 years ago, nobody was willing to pay such a high premium for connectivity.

The project was a failure, but it was an eye-opening experience. I truly started to think about how I wanted to contribute to tech and what was really drawing me to the world of tech as an engineer. This was an era where consumer internet was becoming a popular topic throughout the globe. I remember being worried about the digital divide, and I would come home from college and tell my dad, who was working for *The Los Angeles Times*, that he needed to rethink what he was doing with his career because the internet might put him out of business. He said, 'People have been getting their newspapers at their door since the printing press. The paper is not going anywhere.' Spoiler alert: it is now mostly digital.

When did you become a mother during your tech journey? Did you leave tech to focus on your children? How did having children influence your career?

I have two daughters. One of them is 13 years old, and she is about to move to eighth grade, and the other one is 11, and she is going to be in sixth grade soon. I have a very deep connection with my children, and that definitely has to do with the struggles involved. I have been very vocal about this before, so I will not be afraid to admit that I had a very hard time having children.

I tried for almost eight years to become a mother, and it was a confusing time for me because the doctors were not able to pinpoint why I could not conceive.

After some self-reflection I realized it was likely due to work stress. I did feel stressed every day. I made a decision then, and that was late in 2006, that I would leave tech to focus on having children. I came home and told my husband about the decision. I said that it was for my own sense of mental wellness, and I just wanted to stop what I was doing and focus on having children. Finally, after trying for eight years, I conceived my first daughter and stayed home with her for 18 months—and then I got pregnant with my second daughter soon after. It was a real blessing because I was feeling like my life was tethered to my babies.

At the same time, I started to feel very underutilized professionally. It was during my 26th week of pregnancy when a colleague called and asked me if I would consider joining Yahoo! as a leader. I remember telling her that I was pregnant, and that no matter what, I was going to take a break when week 40 started. I asked her if she was really sure that she still wanted me onboard and she did not hesitate to say yes, because she had children of her own and held a deep respect for the journey and my aspirations to have a family. I accepted the offer, and I was fiercely loyal to Yahoo! because few other companies at that time would hire a pregnant woman.

As for how my relationship has changed while I balance my work and my children, the past 13 years have been eventful. I went from being the one who was there to wipe every nose, wash every bottom, buy every outfit, and be there every step of the way, to being there mainly for the important things. I have since become comfortable with that dynamic because I have a very beautiful relationship with my mother and grand-mother. My own mother was a working mom, so I learned a lot of things on my own. There was a tremendous amount of independence, and it was cultivated in me by my mom who I knew loved me deeply. So, she was the one who taught me how to raise my children and how to give them everything they need to feel loved and successful.

How do you approach being a mom in a way that keeps everything in balance?

I actually do not involve myself with my children in everything, because it does not cultivate the same sense of independence that I want to instill in my daughters. So, when my children say that they really need me there, I will move heaven and Earth to be by their side, but when it comes to the little things, I do not partake in them and leave room for my girls to be creative there. I know they need their space and down-time without me, too, and I personally feel that makes them better people and stronger women.

It has also been really helpful that my husband's mother lives with us, because our kids get to see a multi-generational household where they can spend more time with their grandparents, and that has really helped them to become more empathetic and understanding. It also taught them a lot of life lessons, such as calling 911, dealing with doctors and reacting to emergencies, so they are more acutely aware of all the things that a young adult should learn.

My mother-in-law takes care of a lot of house chores, such as cooking for my children and cleaning out their closets—things I cannot be there for. It actually took me almost two years to become comfortable with that, because I was a brand-new mother and I saw that she was doing everything right and I was not really doing anything at all, so there was a bit of envy, but I eventually realized that my children received a collective benefit from both of us, and it was a very beautiful dynamic. Sharing that stress has really made life easier for me.

Did your husband contribute to the upbringing of your children?

He actually shared a lot of the load of raising our children. I jokingly call him 'Chief Education Officer' because he owned the process for the middle school applications, and he would tell me when he needed my help or my presence while guiding me about all the things he needed me to do. In a way, he was utilizing my skills without putting any additional burden on me, and that was really helpful. We actually have a mindset in our house where we let the expert run the show, so when it comes to athletics, my husband is always there for my children, but when it comes to music, I take time out and sit with my kids and teach them all about playing the piano. I think the best thing to come out of the dynamic is that we have all become better at working as a team. There are times that we have strengths, but sometimes we might all be weak in a specific area, so we band together and work it out collectively. That is one of the biggest and most exhaustive challenges that we face, but it is also where we grow the most as a family.

How have you balanced the pressure of your work with your role as a mother?

There was a point in my career where I decided I was going to stop trying to be like everyone else and just be mean. By this, I do not mean that I was going around insulting other people, it just meant that I was not going to conform to some norms, even if it came at the displeasure of other people. It gave me the space to show who I was, as a mother, as a woman of color and as a public servant. I started being transparent about my right to be a mom at times, and made it clear that it was difficult to be fully available to both my kids as well the organization, as it is actually very challenging to maintain a balance between your work and your role as a mother.

One of the things that I have been really vocal about is our wellness journeys, and it is something that I feel is very important. It could involve locking myself in the bathroom at times to take a bath or dancing at lunchtime because I am feeling low, or even practicing intermittent fasting to deal with all the comfort food I had during COVID. I love giving voice to all of that and giving permission to everyone else to feel it, too, because it is so authentic and it is super helpful in making connection points with my employees, clients, as well as my kids. I am super open with all of that, and it makes me walk a little bit more valiantly through the journey, even when it is hard.

Have you ever felt like an outsider in the tech space?

As a woman of color, I have typically felt a lot more alienated in that regard, but as a woman, there is quite a bit of discrimination on that end, too, especially when we have got a lot of patriarchal, aggressive and dominant stuff going on. Only after dealing with all of that do I ever feel that being a mother is perceived negatively in the tech industry, but I have actually found a lot of alignment with people who have children or are active parents. A lot of us are feeling the pain and pleasure of parenting, so there is a lot of understanding from that end. Personally, being a mother has not particularly held me back from my career, but I do acknowledge that I have not really paid much attention, as the way I have shown up in the room has been profoundly different, so being a parent has had little influence overall.

I have actually felt like an outsider in places outside of the tech sphere, because I have always had limited support as a woman of color. But I have actually learned a lot in my journey to wellness about how to combat that kind of pressure. I have learned to embrace the moments when I feel alone and find alternatives that help me celebrate being alone. I love dancing, but I am 50 years old now, so it does not really feel the same as it did back in high school. But I have found my inspiration to help others through it.

I have started to create those spaces in my work as a leader so I can help others fit in, and that really changes the temperature of the room. I am actually thankful because I find it easier to work with open-stance leaders, and the president of the company fits that role perfectly. His stance and his effective leadership have really made life and work a lot better, and I feel great about the journey, too.

What advice would you give to aspiring women and mothers with a passion for tech?

When I was a little girl, we were having conversations about pro-choice and pro-life, and now, 50 years later, we are still having the same argument. My perspective is that no matter what side of the discussion you're on, it is time to have a different conversation. What I am really asking is that women around the world squeeze a little bit more to ensure that we are pushing a public agenda that creates spaces for us to push private agendas that acknowledge that we are the owners of the largest consumer wallets in the world. That pushes us to continue balancing what we capably do as spouses and mothers and leaders in our households as well, because we can do it. And I am not saying that we drive ourselves to exhaustion—we need to know when to tap out when we need a break, but we also need to figure out all the things we have to explore together.

We women need to tap into our 'extraordinariness',

because this world needs us now to help change the conversations, the temperature in the room and the dynamic as a whole.

What We Learned from Rolanda:

Rolanda has taught us how important it is to create a healthy balance between work and personal life. The way she has raised her children while giving them space to do their own thing, at no additional expense to her career, is definitely something to look up to—and it's important for mothers to realize that they do not need to be there for everything. Rolanda's words are great advice for mothers who feel that they should commit most of their time to their children, and her experiences show how necessary it is to give your children time and distance to build the characteristics they need to be independent and excel through their own efforts.

While mothers feel that they need to be there for everything for their children, the truth is they only need to be there for the important, life-defining moments. Allowing children a degree of independence in minor struggles and maintaining a healthy distance is much more important, as it ensures that your children grow up to be strong and dependent on their own skills. Being there for every step along the way might sound appealing, but it comes at the cost of their development as well as your career. Mothers need to establish a dynamic with their children that balances all these aspects so each individual in the relationship can grow.

Rolanda has also emphasized taking time off and focusing on your own mental health and growth. It is necessary for today's mothers in tech to look after themselves, because they often get caught up between raising children and battling the challenges of work. Taking time out to focus on what gives them peace is crucial, because mental stability is the key component that can help us excel in all other areas.

Similarly, Rolanda explains that asking for help does not make mothers any less significant in their children's lives, and if they can create a dynamic where their children are looked after while still being there for them when it matters, they can find a perfect work-family balance. This also includes discovering each person's strengths in the family and capitalizing on them together, which can definitely alleviate struggles by allowing everyone to contribute where they can.

As some of the other women interviewed also mentioned, Rolanda teaches women and mothers to stand firm in the face of harassment and bullying. Many mothers feel that they have to answer for time they commit to their children, but in reality, this discussion should never be brought up. As women, we need to realize that we do not have any obligation to answer to others about our own well-being. If something needs to be done, it needs to be done and we have to do it. We should not have to make ourselves uncomfortable simply because of other people's conversations and ideas, and we have to discard the guilt that comes with dedicating time to ourselves and our families.

What Tech Leaders Can Learn from Rolanda:

Many tech leaders can learn a lot from Yahoo!'s willingness to host an expecting mother despite knowing that she would inevitably take a break for childbirth, and we hope more companies will follow suit and give space to mothers to prove their worth. Even more than that, however, Rolanda's story emphasizes mental health.

Similar to how Kim also took a stand for mental health, industry leaders need to start discussing the issues which can plague the efficiency and well-being of their employees. Tech leaders need to give employees space to focus on themselves and their external obligations without putting unnecessary pressure on them. The long-standing perception that an employee needs to be at work for their entire shift to be productive no longer holds true in the current era, and tech leaders need to start realizing it. Consideration is the most important concept of the era, and it is time that tech leaders believe it.

Instead of making women and mothers feel like imposters for having additional commitments, tech leaders need to start creating inclusive spaces where mothers can focus on their maternal duties without being guilt-tripped. The president of Rolanda's company is a perfect role model for tech leaders here, as his open stance and effective leadership have created a work environment where women feel welcome.

SEVENTEEN

TURNING LEMONS INTO LIMONCELLO

LIZ'S STORY

> "I want my achievements and my worth as
> a tech worker to be relevant, rather than the
> fact that I'm a female."
>
> —Liz Balmer, Senior Director at a SaaS
> Tech Giant

The struggle to be taken seriously is one many mothers are aware of, but in a world where women's voices are often ignored, we can take inspiration from Liz Balmer. Liz, married with two boys, six and nine, has been working in tech her entire life and is currently a Senior Director of Customer Success, leading a global services sales team at a billion-dollar technology firm. In her words, she manages the department that is "the services-selling arm of customer success."

She started her career as a developer in Big Four

consulting, where she worked with different consulting firms. Liz spent the majority of her career in the Salesforce ecosystem, and then the past five in another large tech company. The challenges Liz faced throughout her career were often associated with people assuming she was younger than her age and correlating that to a lack of experience. After 25 years in tech, she continues to have these challenges today.

* * *

What challenges and barriers have you had to face in your professional journey? How have you overcome them?

I always felt that people thought I was younger than I was and had to prove myself because they had a perception that maybe I was too young. It was more along the lines of 'What is she doing here?' I worked pretty much my entire life with men, but I honestly never felt any kind of gender bias at the time. I think I was just viewed as too inexperienced to be in tech.

That makes me think 'Oh, you do not think I am capable? Fine. I will show you I am. Let us see how it goes.' I have spent most of my career turning these perspectives into a source of strength, and much of that determination comes from viewing it in a positive light.

Do that and your mind will help you find ways to get where you need to be. For me, I just need an opportunity, and I will show results.

Have you had any negative experiences based on how other people have perceived you in the workplace?

I remember one job where I was the most senior person under consideration to lead a team. My manager told me he was not going to give me the manager title until I showed that I could do the job first. He instead gave me a supervisor title, without a compensation change, so I could prove myself. I took the challenge and showed him I was capable and received the title and increased compensation at the end, but I always wondered why I had to do that extra step. When I was the most senior candidate already, having to prove myself was redundant and pointless.

Conversely, when I applied for a position, my mentor reviewed my resume that came across his desk and saw my potential. He decided that I was what he needed for a different role and created a new division in his organization with me in mind. He then hired me to build a new team for that division from the ground up. It was great because I took on the role, and the team under my supervision became the most profitable team within the entire practice, and my mentor was really proud of me. When he went to the next company, he called me again to create a new function and team. In our conversations, he convinced me that I could do the job even if I had little experience running a sales team. When I finally signed, I had a panic attack because sales was newer for me. I was worried I would not be able to tackle it, but my manager was really supportive and told me that he had given me the role because he believed in me. It was actually such a heartwarming experience in itself, to have someone who had faith in me and trusted me to take up the challenge. His trust in me yielded a high performing, business critical global team that supported the company through an IPO and growing suite of products.

With all these negative perceptions, have you ever felt like a feminist?

I am an equalist. I have always worked for men, but now, I actually work for a female leader, and 80 percent of the team is female. One of the men within the team came up to me and said he did not feel like the right fit. I asked him if he felt he was qualified for the job, and he said he was. I asked him if everyone else on the team was qualified too, and he agreed they were. I told him that it is the qualification we really care about, so as long as he was on par in skill, he was equally deserving to be on the team. That has always been my motto, to judge on the basis of skills and qualifications. Applying that concept to myself, I want my achievements and my worth as a tech worker to be relevant rather than the fact that I am a female.

How did you balance the challenges of motherhood with your career?

When I was a new mom, it was much tougher to work in consulting. Despite working from home, I had to send my kids to daycare in the morning, since having screaming children in the background was considered highly unprofessional. I would try to give my kids attention after a nine-to-ten-hour day, but it was difficult working in consulting and not having clear boundaries. There were emails throughout the day, nonstop meetings and work after the kids had gone to bed. My little kids were often sick, so finding child care or negotiating time to look after them were substantial challenges as well. I actually had to hide the needs of my kids at work because there were rumors that I was not as good at my work after becoming a mother. I felt like I had to prove that I was equally great at my job despite having parental duties.

Have you found it challenging to balance your family life with your work life?

I was working with a very difficult client, with 12-hour shifts and onsite work every day when one of my kids was eight months old. One weekend, we were forced to work to support the customers, and I remember sitting at my dining table, talking to my coworkers on the phone while my baby stared at me longingly from his play area. I could see that he wanted me to play with him, and I just had to choose my family over work at that moment. I felt like I couldn't let my teammates down, but I could not ignore my little one's pleading eyes. It's hard to be cheery and engage positively with the kids with work weighing on my mind.

What advice would you give to aspiring mothers in tech?

You do not have to quit completely. There is always a better alternative, and you should look for an organization that is supportive of moms and has a work-life balance culture. After working for a large global consulting firm, I joined a smaller national consulting firm where they had a culture of shutting off email at around 6 pm every day. That meant that emails could only go out during business hours unless there was an emergency. Focusing on family and supporting them was also encouraged, even during business hours, and the only expectation was that we delivered quality work. There are really supportive organizations out there as well, so I hope that moms do not leave tech, but rather look for places that support the working mom life.

I think moms should let go of the belief that they need to do everything at home. One thing I have tried to do is let go of some of the control. I share the work to support the family with my husband now. This helps break some of the home dynamics that have existed for generations where moms typically are responsible for everything at home, even if they are employed. It is actually really innovative because it encourages dads to figure things out, develop better relationships with the kids, and learn new home skills too. I used to map everything out when I would travel for work, but in the past few years, I have left it to my husband to figure it out on his own, in his own way. Doing that can really help to establish more balanced ownership between moms and dads.

What We Learned from Liz:

Liz's story shows that in spite of deterrents and

other people's perceptions, we can still choose to use obstacles as challenges to prove ourselves. Her experiences are not an outlier, as almost 31 percent of women between the ages of 16 and 30 have reportedly faced discrimination while working or seeking employment.[1] In fact, young women are often victims of discrimination because of a lack of experience.

Her experiences also hint at how women are perceived as incapable, and are put through more rigorous testing as compared to their male counterparts, which often means that women have to go the extra mile to match their achievements.

Even so, it is definitely comforting to know that there are still people like Liz's mentor who are willing to give women and mothers the opportunities to prove themselves and show their capabilities. It is important for women to find advocates, not just supporters, that promote and give women opportunities to thrive and prove themselves to be just as capable as their male counterparts.

What Tech Leaders Can Learn from Liz:

As Liz's story clearly shows, mothers should never be in a position where they have to choose work over their children, and tech leaders should take note of that. After all, it is incredibly painful to commit to work at the expense of time you should be spending with your family.

Also, while women contribute just as much as men do in the tech industry, there are still times when their authority or ability are overly questioned. More troubling is that sometimes, even well-meaning leaders do not realize when they are being discriminatory or sexist, which puts women working for them in a difficult place. Instead of putting this burden on other team members, male tech leaders need to exercise self-awareness and create spaces where these conversations can be had openly and safely.

Finally, Liz's story shows that more and more companies around the globe are shifting their focus to the needs of their employees, particularly in light of the COVID pandemic. The dynamic has shifted towards a workplace culture that intermingles with family life. While sexism and discrimination still exist within the industry, companies are slowly shifting towards more inclusive workplaces that allow individuals to address external responsibilities at the same time. For tech leaders who want their companies to be as competitive as they can be, they will need to follow suit in order to keep up.

PART 5

CONCLUSIONS AND FUTURE CHALLENGES

EIGHTEEN

THRIVING AS A MOTHER IN TECHNOLOGY

The tech industry is an exciting and complicated jungle that we are passionate about and proud to be part of. It is filled with many, often opposing challenges, deadlines, people and ideas. We have experienced firsthand that it is much more difficult for a woman or mother to make her space in tech. These limitations and setbacks can be deal-breakers for even the most determined individuals. Knowing how to navigate a complicated environment that is often plagued with discrimination and prejudice is essential to surviving in the tech sphere.

When we developed the concept behind this book, one of the main motives was to give mothers the necessary tools and motivation they needed to make a name for themselves in tech, and this chapter serves to assist them in that journey.

As a mother or woman aspiring to succeed in tech, one of the most important concepts to internalize is that tech is not a man's world. Often, we tend to believe that the typically prevalent "boys' club" culture and disparities between male and female employees make tech a field suitable only for men. That is a common misunderstanding that only serves to further fuel the divide. The very first step towards prosperity in tech is to eliminate the mindset that tech is a man's world, not just on an individual level but also on a collective level. However, the process is two-fold.

There needs to be significant representation of women in tech to truly dispute this mindset on a collective level. Therefore, every woman and mother with a passion for tech needs to instill in themselves the belief that tech is everyone's world. As many of the women in this book have said, it is all about the mindset. To truly succeed in tech, we have to embrace that we are just as important as our male counterparts. Women and mothers belong in tech, and owning this is crucial towards motivating yourself, your sisters and your friends towards success.

While the mindset of success is incredibly important in the tech sector, it is essential to realize that it's challenging to be consistently and continuously positive, especially considering the amount of discrimination women and mothers face in the industry. It is perfectly valid to feel imposter syndrome at times simply because tech can often feel unwelcoming or unsupportive of women and mothers. However, it is equally essential to understand that combatting imposter syndrome is only possible if we step out and prove our worth.

When we are constantly providing results, it is impossible for anyone to point accusatory fingers at us and it grants us the confidence to feel comfortable in tech. If women and mothers own their place in tech and give their best, they can defeat imposter syndrome and truly feel that they and their work have an impact. Not only will this mindset help women and mothers achieve their dreams, it will also help increase the representation of women across all teams working in the tech sphere. This inclusiveness will motivate and encourage other women to step up as well, as more female role models in tech will help other women feel they also belong in that world.

On the topic of battling imposter syndrome, it is just as crucial for women and mothers to realize that they need to stand up for and take care of themselves. Only by being vocal about their rights and require-ments can women aim to change the status quo in their favor. As many women in our survey have pointed out, nobody can speak up for you if you are not willing to speak up for yourself. There will often be situations where you are your only advocate, and in those moments, it is essential to step up and make your voice heard. In the face of harassment and discrimination, it is crucial to take a stand, as allowing it to continue only allows the discrimination to increase in severity.

When we talk about taking a stand, it applies equally to yourself and others. Standing up for other minorities and setting a precedent where women's rights are respected and their voices are heard is essential. Women making a name for themselves in tech need a support system, and women in higher positions can often advocate for them and ease their journeys.

Senior women standing up for other women can play an important role in their success. It is also equally important for women starting off or in junior positions to seek role models to advocate for them. We advise all women with a passion in tech to find role models and mentors who can guide them along the journey and teach them how to navigate the challenges and struggles of a tech career. Having someone who can speak on your behalf can make your presence known even when you are not there, and that can be vital towards establishing your worth in an organization.

Despite how strong women have to be in the face of oppression in the tech industry, it is essential to accept that sometimes, the burdens can pile up, especially considering the pressures of family life that come with being a mother. Mothers need to accept that it's perfectly reasonable to want time off from work and to take a break, and in no way does this make them weak or incapable. Everyone, irrespective of gender, can succumb to the constant pressure of the tech world, so we need to accept that it is okay to take time for yourself or for your family. It is also equally acceptable to prefer comfort over societal pressure, so presenting yourself in a certain way just to appeal to the ideals of society and the workplace, at the cost of your personal well-being, is never the right choice.

Mothers and women need to create their own comfortable space and ignore or decline situations that might compromise it, as doing so can result in low productivity, unhappiness, anxiety and feelings of disempowerment. Psychological security is an essential component of success, and is the best way for women to excel regardless of their position.

While we have discussed how important it is to be vocal about your needs as a mother, it is equally important to be vocal about your achievements. Many women, including the mothers in our survey, have made massive strides in technology, going as far as reshaping the entire industry into a more productive and holistic space. However, as we have discussed before, when we look at individual achievements, there are very few women who stand out. While this disparity has been influenced by sexism, it is necessary to understand that these contributions can only be realized by speaking about them.

We advise mothers and women to be vocal about their work as their careers grow. Talking about what they have learned, their common mistakes and the impact they are having are important steps to normalizing and applauding women for building their careers. Building connections this way with one's seniors is also crucial in building a reputation, and with a good reputation comes strong representation. Not everyone is an extrovert, but practicing being vocal and having a presence in the organization is crucial so when management is deciding who to give leadership roles to, your name remains at the forefront. Women can be amazing at their profession, but they will not achieve success until they display their work for others to see. Technology is a race—so no matter how good you are at running, you still have to join the race to win.

Perhaps the most important advice we can give to mothers in tech is to focus on work-life balance while being okay with knowing that it will never be perfectly balanced. Doing justice to both aspects is important but is different for everyone. Some parents might not be obligated to commit to their children as heavily, while others might feel that their work is affecting family life and will need to cut down on their work-related responsibilities. Since each person has a different balance to maintain, there is no definitive way forward; it is something everyone has to figure out for themselves. The best way to know that you have achieved balance is when you are comfortable with the arrangement you have developed. Always remember: doing justice to your work and your family offers a peace that is unmatchable, and once you achieve it, everything simply falls into place.

As a closing note, we wish all mothers—those starting off and those who are already established in tech—strength in their journeys, and we hope they achieve all that they wish. As working mothers ourselves, we want to see more mothers take up the mantle and prove their worth. We want to see greater representation of mothers in tech, and we want to be there with them when they inevitably change the tech world for the better.

NINETEEN

THE IMPACT OF COVID-19 ON WORKING MOMS

> "According to the Center for Work-Life Policy, 56 percent of women leave the tech industry mid-career, more than double the rate of men. The struggle is real, but the expertise tech moms bring to the industry is critical."
>
> —Chas Larios, Tech Mom Collective

Early data shows that the number in the statistic above is climbing in the wrong direction. There is no doubt that the COVID-19 pandemic, which significantly influenced the world economy and adversely impacted a majority of the world population including mothers in tech, has made things worse—though it is not the only thing to blame. With the impact COVID-19, a majority of nonessential industries were forced to shut down to contain the spread of the virus. While essential services remained active, the model of operations for businesses everywhere switched to a more digitized approach that limited in-person interaction. The result was a switch towards online operations, where remote working became the norm. In the tech sector, there were both advantages and disadvantages to the abrupt change. Though most businesses in tech were quick to adapt to the new model since the infrastructure already existed to support it, there was also reduced demand for tech services, primarily in nonessential sectors, and businesses transitioned from making profits to being forced to survive. The result was the closure and downsizing of businesses, particularly for those that relied heavily on in-person meetings and services, which resulted in many businesses closing, leaving many individuals jobless.

According to a survey by the National Women's Law Center, 2.2 million women left the labor force between February and October 2020.[1] While one might assume that the shift to a work-from-home model might have been helpful to working mothers by allowing them easier access to their families and more time at home, the overall impact has actually proven negative. According to the US Census Bureau, 3.5 million mothers with school-age children left their jobs in the spring of 2020, up from the previous year.[2]

There were many challenges that mothers faced during the pandemic, like the closure of daycare centers due to quarantine restrictions. Moms were forced to cut down on their hours working so that they could also homeschool their children, make lunches and change diapers. Quarantining to isolate from the potential virus spread presented mothers with the dilemma of choosing to look after their children at the expense of their career and financial stability, or focusing on their careers and neglecting their children.

The pandemic also limited the opportunities women had to use external help. Before the pandemic, most mothers could rely on their children being in school, leveraging babysitters, daycare facilities and family members to help manage the needs of their children. However, during the pandemic, social distancing meant that it was deemed medically unsafe to co-mingle with other households and therefore families did most of life's activities under one roof. While there are supportive fathers and male partners that share in the burden of being the children's academic educator, physical education instructor and chef, while attempting to host video conference calls and produce quality work, the preliminary studies have shown that it was women who owned these primary duties. The 3.5 million women from the US Census study is inclusive of mothers who were laid off after being seen as "under-productive."

According to a statistic by OECD, mothers had to work four additional hours a week than men to maintain their jobs during the pandemic, which indicates how much added pressure they faced.[3] Part of this added stress came from mothers needing to manage the burden of education, as school closures limited their children's educational opportunities. With these combined factors, it is understandable that mental health was negatively impacted. In fact, according to a 2020 report by TUC, 90 percent of working mothers reported a decline in their mental health during the pandemic.[4] Feelings of inadequacy, worthlessness or exhaustion became prevalent, all the while also negatively impacting their social position in both their personal and professional lives.

Overall, the pandemic was a bane for mothers' careers worldwide, and working moms in the technology industry were not spared. Despite these challenges, we know that mothers are resilient. According to a survey by McKinsey, mothers have more determination and solid career goals than women without children, and if there's one thing we know, it is that no matter what the challenge, mothers are capable of tackling it.[5] In spite of whatever the world throws at mothers, mothers will find a way to excel.

RESOURCES TO FACILITATE YOUR JOURNEY

It can often be challenging for mothers in tech to navigate the difficulties and challenges they face, particularly for mothers who are new to the sector. For those mothers, this section is designed to be a resource and guide, offering the tools necessary to ensure their success.

As the authors of this book are mothers themselves, we'll start off by telling you our personal strategies. In short, this is how we do it:

1. Find Sponsors

The most important step in your journey is finding a sponsor. This can be difficult in workplaces with limited representation of women, but we have learned that there are always a handful of individuals who will stand up for you and guide you. A sponsor can also give you a much-needed confidence boost as you make your own way within an organization. Make friends and acquaintances, and ask for help and advice if you need it. Once you find your sponsors who will stand up for your rights and support you, your journey will become a lot easier.

2. Plan, Plan and Plan

Being prepared can make even the most challenging of tasks much easier. That is why it is important to plan your day and work while still leaving enough space for any urgent or emergency occurrences. For mothers who have to balance work and family, having a proper timetable allows them the ability to dedicate time and attention to all necessary aspects of their lives. These timetables can include activities such as meal planning, picking up children from school or even taking time off to relax.

An organized life is a peaceful life, and while we understand that things might not always go as expected, knowing how to manage your time is a key skill that can help you focus on the things that matter. Having your responsibilities listed can also help ensure that you are not skipping out on the important things, and that balance can help you excel in all areas.

3. Carve Out "You" Time

When planning your day, always focus on your strengths and comforts. For example, Sabina is a night owl and prefers working when everyone is asleep, as it allows her to dedicate her focus completely on her work. Emilia, on the other hand, is an early riser and prefers to dedicate the early hours to her work. Both of us manage our time in a way that allows us to give attention to various parts of our lives while still taking time for ourselves. Taking that time for yourself is crucial, as it enables the mental peace and stability you will need to maintain balance in your life.

Podcasts to Inspire You:

There are multiple podcasts that you can listen to while driving or doing your daily activities that can offer helpful insight and tips to assist you in your journey. Here are a few of our favorites:

1. *The Women in Tech Show* by Edaena Salinas
2. *Women Tech Charge* by Dr. Anne-Marie Imafidon
3. *Witty: Women in Tech Talk to Yaz* by Yasmin Alameddine
4. *Big Careers, Small Children* by Leaders Plus
5. *Women in Tech* by Espree Devora

6. *Her STEM Story* by Prasha Dutra

7. *Women in Tech with Ariana* by Ariana the Techie

Top Guidebooks:

1. *Brotopia: Breaking Up the Boys' Club of Silicon Valley* by Emily Chang

2. *Reset: My Fight for Inclusion and Lasting Change* by Ellen Pao

3. *Girl Code: Unlocking the Secrets to Success, Sanity, and Happiness for the Female Entrepreneur* by Cara Alwill Leyba

4. *Women in Tech: Take Your Career to the Next Level with Practical Advice and Inspiring Stories* by Tarah Wheeler

5. *The Adventures of Women in Tech: How We Got Here and Why We Stay* by Alana Karen

6. *Lean In* by Sheryl Sandberg

7. *How Women Rise* by Sally Helgesen and Marshall Goldsmith

8. *Untamed* by Glennon Doyle

9. *Better Allies: Everyday Actions to Create Inclusive, Engaging Workplaces* by Karen Caitlin

Webinars:

1. *BrightTALK: BrightTALK* offers an up-to-date list of

all webinars about women and mothers in technology. It is a perfect resource for moms looking for discussions by female tech leaders and mothers of influence.

2. *WeAreTechWomen*: This resource offers mothers a list of recorded webinars that can help them expand upon their career and improve their own skills. It further offers the latest updates on upcoming webinars.

3. *The WIT Network*: *The WIT Network* offers a range of webinars on various topics and discusses common challenges mothers face in their careers and in the tech world while providing suitable resolutions.

4. *Build an "And" Culture and Thrive*: This webinar by Linux Academy provides women and mothers solutions on how to tackle career challenges and effectively thrive within their jobs.

Insightful Articles:

1. "Google advised mental health care when workers complained about racism and sexism" (*NBC News*)
2. "Learning from Women Who've Made It to the Top in Tech" (*BCG*)
3. "Why do women in STEM feel the need to 'prove themselves' in the workplace?" (*Silicon Republic*)

4. "Women in tech statistics: The hard truths of an uphill battle" (*CIO*)

5. "The 5 Biases Pushing Women Out of STEM" (*Harvard Business Review*)

6. "TrustRadius 2021 Women in Tech Report" (*TrustRadius*)

7. "Moving Goalposts: Kimberly Lowe-Williams on Black Leadership and the Importance of Paying It Forward at Salesforce" (*AfroTech*)

8. "How to Fix Women's Jobs During the COVID-19 Pandemic" (*TIME*)

9. "Why So Many Companies' Diversity Numbers Fall Flat" (*Fast Company*)

10. "The Look Of Power: How Women Have Dressed For Success" (*NPR*)

11. "Power Dressing" (*Wikipedia*)

12. "The Secret History of Women in Coding" (*New York Times*)

13. "Why so Few Women in Computer Science? A Look into Stereotyping & College Curriculum" (*Medium*)

14. "Managers: This Is How To Succeed Where Google Failed In Supporting Working Parents" (*Forbes*)

15. "Equality at work is good for biz" (*LinkedIn*)

Support Organizations:

The Mom Project: *The Mom Project* helps working mothers discover career opportunities. It also assists in integration within the workplace as well as finding mentorship for mothers to unlock their true potential.

Women In Technology (*WIT*): *WIT* is committed to helping working mothers through inspiring stories and support groups, with a focus on inculcating the necessary skills to help them achieve a successful future in tech.

Maven: *Maven* is a digital health platform that works with health plans and employers to offer virtual services for women's health and family health.

Moms Rising: *Moms Rising* raises the voices of mothers across the US while offering advice on how to tackle persistent workplace issues, with guides to help mothers tackle challenges systemically to ensure they rise above discriminatory practices.

Moms with Careers Making it All Work: This Facebook group offers mothers advice on challenges they may face in their careers and personal lives. With the option of asking for advice and guidance, the group lets mothers be vocal about the challenges they face while receiving assistance from other working mothers. This support group is also helpful for forming connections with mentors and discovering job opportunities.

MotherCoders: *MotherCoders* helps moms connect to job opportunities while offering additional support such as on-site child care. The training modules help instill the skills necessary to integrate into the fast-paced tech industry. For mothers with financial difficulties, child care commitments or personal struggles, they offer support on multiple fronts.

Ada's List: An email-based support group that allows mothers to seek advice and guidance for their tech careers that helps mothers discover career opportunities and connect with other women to build mentor relationships. *Ada's List* also has an annual conference that can help you connect with other members in person.

Change Catalyst: *Change Catalyst* helps build inclusive environments in tech through startup advice and events around the globe. It is perfect for mothers looking to connect with other moms, and helps create more mother-friendly spaces in workplaces.

Girls In Tech: *Girls In Tech* offers several courses and programs teaching mothers necessary workplace skills, with an additional focus on techniques to help them rise through the organization's hierarchy. *Girls In Tech* is a perfect resource to help mothers and tech leaders develop inclusive environments in the workplace, and to create environments where everyone feels welcome.

iRelaunch: iRelaunch helps mothers who have taken breaks from their careers reintegrate into tech. It can assist women who have gaps in their CVs reenter tech and continue on their journey who might otherwise be sidelined.

Get Our Complete Guide:

Our complete guide for mothers is available on our website www.pressingonbook.com and covers all the resources and tips that can help you along your journey as a mother in tech.

Connect With Us On LinkedIn

Sabina M. Pons (https://www.linkedin.com/in/sabina-pons)
Emilia D'Anzica (https://www.linkedin.com/in/emiliadanzica)

ACKNOWLEDGMENTS

Emilia:

To my family. We have been through the loss of Emilio, my brother who was only 11 when a drunk driver ended his life. I didn't have the opportunity to meet you but I carry on your love of life and wear your Montreal Canadiens Jersey with pride. Mamma e Papa, Margherita and Antonio, we have been through so much. Regardless of the pain that we have endured, we five remaining siblings, Josephine, Guilia, Maria, Sammy, remain bonded and thank you for the courage you share with us each day.

To my partner in life, Jason. Thank you for supporting my decisions even when you didn't agree... my willingness to choose the harder path. You are the calm for my storm. The moment we met, I knew I didn't want to ever let you go.

To Sabina! I am thankful for our deep friendship. We inspire and elevate each other. I love you. Karey, we were supposed to open a law practice together but life took us on different paths. Regardless, you remain my best friend.

Sabina:

To my husband, Shawn, you are the embodiment of what it means to be a true partner. You encouraged me to pursue a master's degree while we simultaneously held rigorous careers, supported us in writing this book, and have been there for me for all of the things in between, like caring for me through mental health challenges and physical recoveries. You put up with my juxtaposition of being a dreamer and a worrier ("where's the crib going to go?!") and you do it with a warm hug and a smile. Thank you. I love you.

To my parents, Peggy, Mark and Steve, and to my siblings, Krista, Collin and Brooke—your patience, love, and laughter have always been my light. I love and appreciate you.

To Oma & Opa, I think that you knew that I'd write a book before I knew that I'd write a book. Thank you for enriching my childhood. I love and miss you.

To Katie B., April, Maria, Katie A., Taylor and Val. You are my oldest and truest friends. Thank you for your support through it all. Muchos besos.

Emilia, thank you for stomping out my imposter

syndrome and for enabling my career. You are the big sister that I always wanted.

Emilia & Sabina:

To the brilliant women who vulnerably allowed us to interview them—Tracy, Jennie, Liz, Angela, Rolanda, Kim, Jodie, Jackie, Amanda, Priya, Lauren and Christina—thank you!

To the 318 women that participated in our first-of-its-kind research study, thank you for your time and openness.

To Taylor, for your creativity and quick thinking—you are awesome. To Anna, Kaitlin, Ryan and the rest of the team at Launch Pad— thank you for seeing our potential. To Sarah, thank you for putting up with our hectic schedules and for making it all run smoothly. To Kristin—we've only just begun what will be a power-house of a partnership. Thank you for seeing our vision and sharing it with the public.

Thank you to The Authoresses for your key learnings and willingness to lift one another up in our writing journeys. We are grateful for the inspirational examples put forth by you and other female authors. Thank you to Alana Karen, Maria Ross, Jeanne Bliss, Sarah E. Brown and Donna Weber for sharing your book writing tips with us, even when your own schedules were hectic.

To Nick, Scott, Jacco, Chris, Dominique and Sean —you've believed in us, mentored us and invested in us. The world needs more leaders in tech like you. Thank you for your allyship and sponsorship.

NOTES

1. Women in Technology

1. Wikipedia. *1980 Irpinia Earthquake.*
2. Vogel, Joseph. *The Nation That Janet Jackson Built.*
3. University of South Florida. *Generational Differences Chart.*
4. Jacobson, Lindsey. *Equal Pay Day 2018: Stunning studies reveal highest paid women face the greatest gender wage gap.*

2. The Book's Mission

1. American Association of University Women. *The STEM Gap.*
2. VanHack. *Women in Tech Statistics: What the Numbers Tell Us.*
3. Parmar, Belinda. *Are Men Just Better at Technology?*
4. Mims, Christopher. *The First Women in Tech didn't Leave— Men Pushed Them Out.*
5. PwC. *Women in Tech—Time to Close the Gender Gap.*
6. Entelo. *Quantifying the Gender Gap in Technology.*
7. HackerRank. *2018 Women in Tech Report.*
8. Schembari, Mariam. *10 of the Best Companies with Paid Maternity Leave.*
9. Wakabayashi, Daisuke & Benner, Katie. *How Google Protected Andy Rubin, the 'Father of Android'.*

3. A Note On Breaking Imposter Syndrome

1. Tulshyan, Ruchika & Burey, Jodi Ann. *Stop Telling Women They Have Imposter Syndrome.*

2.   Hendricks, Gay. *The Big Leap: Conquer Your Hidden Fear and Take Life to the Next Level.*

3.   Sakulku, Jaruwan & Alexander, James. *The Impostor Phenomenon.* International Journal of Behavioral Science, 2011, Vol. 6, No. 1, 75-97.

4.   Salter, Chuck. *Failure Doesn't Suck.*

4. Finding Joy At Home and Work

1.   Cooper, Marianne. *Mothers' Careers Are at an Extraordinary Risk Right Now.*

2.   McKinsey & Company. *Women in the Workplace 2020.*

3.   Elliot, Brian. *Hybrid rules: The emerging playbook for flexible work.*

4.   Lublin, Joann S. *How Women Can Ditch the Guilt When Juggling Careers and Family.*

5. Claiming Your Own Path

1.   Richter, Felix. *Women's Representation in Big Tech.*

2.   Daley, Sam. *Women in Tech Statistics Show the Industry Has a Long Way to Go.*

6. Listening to Your Inner Voice

1.   Kokil, Sneha. *Closing the gender gap in today's tech industry.*

2.   Light, Paulette. *Why 43% of Women With Children Leave Their Jobs, and How to Get Them Back.*

7. Embracing Mentorship and Sponsorship

1. Franceschi-Bicchierai, Lorenzo. *I'm Not Returning to Google After Maternity Leave, and Here Is Why.*
2. Hale Alter, Cara. *LinkedIn Home.*
3. Ibarra, Herminia. *A Lack of Sponsorship Is Keeping Women from Advancing into Leadership.*
4. Hewlett, Sylvia Ann. *The Real Benefit of Finding a Sponsor.*
5. TrustRadius. *TrustRadius 2021 Women in Tech Report.*

8. Rising Above Your Challenges

1. Society for Human Resource Management. *Preventing Unlawful Workplace Retaliation in California.*
2. U.S. Equal Employment Opportunity Commission. *Retaliation.*
3. Paul, Kari. *She sued for pregnancy discrimination. Now she's battling Google's army of lawyers.*
4. Golshan, Tara. *Study finds 75 percent of workplace harassment victims experienced retaliation when they spoke up.*
5. Bianchi, Suzanne M., Robinson, John P. & Milkie, Melissa A. *Changing Rhythms of Family Life.*
6. US Department of Labor, *12 Stats about Working Women.*

14. Motherhood Made My Success

1. Lien, Tracy. *Tech's glass ceiling nearly four times harder for Asian Americans to crack.*
2. Geuss, Megan. *Former Facebook employee retains Ellen Pao's lawyer in new discrimination case.*

17. Turning Lemons into Limoncello

1. Siddique, Haroon. *Workplace gender discrimination remains rife, survey finds.*

19. The Impact of COVID-19 on Working Moms

1. Ewing-Nelson, Claire. National Women's Law Center. *Nearly 2.2 Million Women Have Left the Labor Force Since February.*
2. RadioEd. *The "Motherhood Penalty": COVID's Impact on Working Women.*
3. Organization for Economic Co-operation and Development. *Employment: Time spent in paid and unpaid work, by sex.*
4. TUC. *Working mums: Paying the price.*
5. Huang, Jess, Krivkovich, Alexis, Rambachan, Ishanaa &Yee, Lareina. McKinsey & Company. *For mothers in the workplace, a year (and counting) like no other.*

ABOUT THE AUTHORS

 Emilia M. D'Anzica is the Founder and Managing Director of Growth Molecules, a management consulting firm focused on customer support and success. The company's mission is to help organizations increase profit while maximizing value to customers. Emilia is also on several advisory boards globally and an active contributor of the Forbes Council.

As an early employee at several successful companies, Emilia amassed more than 20 years of customer experience in roles as Vice President of Customer Engagement at WalkMe, Director of Client Service Operations at BrightEdge and Director of Customer Success at Jobvite.

Emilia holds a BA from the University of British Columbia and an MBA with Honors from Saint Mary's College of California. She is PMP and Scrum certified.

Emilia resides in the San Francisco Bay Area with her three children and partner. *Pressing ON as a Tech Mom* is her first book.

Sabina M. Pons is an award-winning management consultant with a focus on driving revenue protection and growth for technology companies. In her 20+ year career, she has led global corporate teams, managed multi-million-dollar P&Ls and built teams from the ground up, all with the mission of inspiring customer centricity with a human-first philosophy.

With a master's degree in Communication, Leadership and Organizational Behavior, Sabina is passionate about igniting corporate transformational change. She sits on several boards, participates in many mentorship programs and is the proud new owner of a first-degree black belt in Taekwondo. She resides in Orange County in Southern California with her husband, two young children and goldendoodle Riley. *Pressing ON as a Tech Mom* is her first book.

Made in United States
Orlando, FL
26 May 2022

18191616R00159